Plasma Processes for Renewable Energy Technologies

Plasma Processes for Renewable Energy Technologies

Special Issue Editor

Masaaki Okubo

MDPI • Basel • Beijing • Wuhan • Barcelona • Belgrade

MDPI

Special Issue Editor
Masaaki Okubo
Osaka Prefecture University
Japan

Editorial Office
MDPI
St. Alban-Anlage 66
4052 Basel, Switzerland

This is a reprint of articles from the Special Issue published online in the open access journal *Energies* (ISSN 1996-1073) 2019 (available at: https://www.mdpi.com/journal/energies/special_issues/Plasma_Processes_for_Renewable_Energy_Technologies)

For citation purposes, cite each article independently as indicated on the article page online and as indicated below:

LastName, A.A.; LastName, B.B.; LastName, C.C. Article Title. *Journal Name* **Year**, *Article Number*, Page Range.

ISBN 978-3-03921-972-8 (Pbk)
ISBN 978-3-03921-973-5 (PDF)

Contents

About the Special Issue Editor

Masaaki Okubo is a Professor in the Department of Mechanical Engineering at Osaka Prefecture University. His current research interests include environmental applications of nonthermal plasmas, particularly nanoparticle control, electrostatic precipitator, aftertreatment for super-clean diesel engines and combustors, and surface treatment for materials and its biomedical applications. Dr. Okubo works in the multidisciplinary areas of electrical, chemical, and mechanical engineering. He has authored more than 150 peer-reviewed papers in scientific journals and authored 12 books. Masaaki Okubo received his B. Eng., M. Eng., and Ph.D. degrees from the Department of Mechanical Engineering, Tokyo Institute of Technology, Tokyo, Japan, in 1985, 1987, and 1990, respectively. He worked as an invited Professor of Tohoku University in 2015. He is a Fellow of the Japan Society of Mechanical Engineering and received the Environmental Engineering Award at the Environmental Engineering Division in 2013. He was also a chairman of the Electrostatic Process Committee and is an associate editor of the IEEE Industry Application Society, the Osaka region chairman of the Institute of Electrostatics Japan, and an editorial board member of the Journal of Electrostatics published by Elsevier.

Preface to "Plasma Processes for Renewable Energy Technologies"

One of the plasma industry applications expected to develop greatly in the future is environmental protection technology. Although some environmental improvement systems, such as odor control machines or deodorizers, indoor air cleaners, electrostatic precipitators, etc., have already been put to practical use, there are new needs in a wide range of fields such as the removal of atmospheric pollutants, cleaning of exhaust gas from combustion equipment, purification of liquid pollutants, decomposition of volatile organic compounds, promotion of combustion by plasma, and purification of the indoor environment. It is hoped that plasma technology will be applied and developed through various approaches in the future as energy-saving environmental improvement technology for combustion equipment. Accordingly, a Special Issue of the journal Energies on plasma processes for renewable energy technologies was planned in response to a request by MDPI. To publish papers that widely disseminate the role and potential of plasma from the keyword "Environmental protection", we requested works pertaining to cutting-edge research contents from experts who have been engaged in research and development for many years in the field of environment plasma, and papers in this field were solicited. In this issue, we could publish papers on environmental plasma technologies that can effectively utilize renewable electric energy sources. However, any latest research results on plasma environmental improvement processes were accepted for publication. As a result, eight high-level papers were collected and peer-reviewed. An excellent description of an atmospheric cleaning process and environmental plasma applications was presented. This book is a compilation of these articles of this Special Issue. The explanations and editorial details of each original paper are explained in the "Editorial". It should be clearly stated that the contents of this book include some of the remarkable achievements related to environmental plasma technology but do not encompass the full scope of research. I hope that this book will contribute to the development of environmental plasma application technologies and facilitate technology development to restore a blue sky and get rid of the dirty sky caused by air pollution.

Masaaki Okubo
Special Issue Editor

energies

MDPI

Editorial

Special Issue on Plasma Processes for Renewable Energy Technologies

Masaaki Okubo

Department of Mechanical Engineering, Graduate School of Engineering, Osaka Prefecture University, 1-1 Gakuen-cho, Naka-ku, Sakai 599-8531, Japan; mokubo@me.osakafu-u.ac.jp; Tel.: +81-72-254-9230; Fax: +81-72-254-9233

Received: 14 November 2019; Accepted: 15 November 2019; Published: 20 November 2019

1. Introduction

The use of renewable energy is an effective solution to mitigate global warming. Environmental plasma processing is also an effective means to mitigate global environmental hazards arising from the emission of nitrogen oxides, (NO_x), sulfur oxides (SO_x), particulate matter (PM), volatile organic compounds (VOC), and carbon dioxide (CO_2) into the atmosphere. By combining both technologies, we can develop an extremely effective environmental improvement technology. Nuclear energy used for power generation is another effective source for the generation of discharge plasmas. Accordingly, a special issue of the journal *Energies* on plasma processes for renewable energy technologies was planned. In this issue, we focused on environment plasma technologies that can effectively utilize renewable electric energy sources, such as photovoltaic power generation, biofuel power generation, and wind turbine power generation. However, any latest research results on plasma environmental improvement processes were welcome for submission. We were looking for studies on the following technical subjects, among others, in which plasma can either use renewable energy sources or be used for renewable energy technologies:

- Plasma decomposition technology of harmful gases, such as the plasma denitrification method;
- Plasma removal technology of harmful particles from combustion machines, such as electrostatic precipitation;
- Plasma decomposition technology of harmful substances in liquid, such as gas–liquid interfacial plasma;
- Plasma-enhanced flow induction and heat transfer enhancement technologies, such as ionic wind device and plasma actuator;
- Plasma-enhanced combustion and fuel reforming;
- Other environmental plasma technologies.

The keywords are as follows: nonthermal plasma, plasma denitrification, electrostatic precipitator, gas–liquid interfacial plasma, ionic wind, plasma actuator, plasma-enhanced combustion, and fuel reforming.

2. A Short Review of the Contributions to This Issue

The contributions to the special issue are reviewed briefly as follows.

Liu et al. [1] contributed a paper entitled "Experimental and Numerical Investigations of Plasma Ignition Characteristics in Gas Turbine Combustors". This study reported a reliable ignition, which is critical for improving the operating performance of modern gas turbine combustors. Recently, plasma-assisted ignition has attracted interest to realize combustion improvement in internal combustion engines. Based on an optical experiment, the plasma jet flow feature during discharge was analyzed. Then, a detailed numerical study was conducted to investigate the effects of different plasma parameters on the ignition enhancement of a combustor used in gas turbines. The results showed that plasma has a good ability to expand the ignition limit and decrease the minimum ignition energy. For the studied plasma ignitor, the initial discharge kernel was not a sphere, but a jet flow cone with a length of approximately 30 mm. Furthermore, the numerical comparisons indicated that the additions of plasma active species and the increases in the initial energy, plasma jet flow length, and discharge frequency can benefit the acceleration of kernel growth and flame propagation via thermal, kinetic, and transport pathways. The result is very effective for the improvement of combustion in internal combustion engines and high-performance combustion.

Ahn et al. [2] contributed a paper entitled "Control Strategy for Power Conversion Systems in Plasma Generators with High Power Quality and Efficiency Considering Entire Load Conditions". In this paper, a control method for the power conversion system (PCS) of plasma generators connected with a plasma chamber was presented. The PCS generated the plasma by applying a high-magnitude and high-frequency voltage to the injected gases in the chamber. With regard to the PCS, the injected gases in the chamber were equivalent to the resistive impedance, and the equivalent impedance had a wide variable range, according to the gas pressure, the amount of injected gases, and the ignition state of gases in the chamber. In other words, the PCS for plasma generators should operate over a wide load range. Therefore, a control method for the PCS in plasma generators was proposed to ensure stable and efficient operation over a wide load range. In addition, the validity of the proposed control method was verified via simulation and experimental results based on an actual plasma chamber.

Tamošiūnas et al. [3] contributed a paper entitled "Gasification of Waste Cooking Oil to Syngas by Thermal Arc Plasma". The objective of this experimental study was to conduct experiments gasifying waste cooking oil (WCO) to syngas. WCO can be used as an alternative potential feedstock for syngas production. The WCO was characterized to examine its properties and composition in the conversion process. The WCO gasification system was quantified in terms of the produced gas concentration, H_2/CO ratio, lower heating value (LHV), carbon conversion efficiency (CCE), energy conversion efficiency (ECE), specific energy requirements (SER), and the tar content in the syngas. The best gasification process efficiency was obtained at the gasifying-agent-to-feedstock (S/WCO) ratio of 2.33. At this ratio, the highest concentrations of hydrogen and carbon monoxide, the H_2/CO ratio, the LHV, the CCE, the ECE, the SER, and the tar content were 47.9%, 22.42%, 2.14, 12.7 MJ/Nm^3, 41.3%, 85.42%, 196.2 kJ/mol (or 1.8 kWh/kg), and 0.18 g/Nm^3, respectively. The authors concluded that the thermal arc–plasma method used in this study can be effectively used for the gasification of WCO to high-quality syngas with a low content of tars.

Yamasaki et al. [4] contributed a paper entitled "Plasma–Chemical Hybrid NO$_x$ Removal in Flue Gas from Semiconductor Manufacturing Industries Using a Blade-Dielectric Barrier-Type Plasma Reactor". A combustion abatement system is used to treat perfluorinated compounds (PFCs), which are used in the semiconductor manufacturing system. NO$_x$ is emitted in the flue gas from semiconductor manufacturing plants as a byproduct of the combustion for the abatement of PFCs. To treat NO$_x$ emission, a combined process consisting of a dry plasma process using nonthermal plasma and a wet chemical process using a wet scrubber was performed. For the dry plasma process, a dielectric barrier discharge plasma was applied using a blade-barrier electrode. Two oxidation methods, direct and indirect, were compared in terms of NO oxidation efficiency. For the wet chemical process, sodium sulfide (Na$_2$S) was used as a reducing agent for NO$_2$. Experiments were conducted by varying the gas flow rate and input power to the plasma reactor, using NO diluted in air to a level of 300 ppm to simulate exhaust gas from semiconductor manufacturing. The results demonstrated that the proposed combined process is promising for treating NO$_x$ emissions from the semiconductor manufacturing industry.

Kawada et al. [5] contributed a paper entitled "Development of an Electrostatic Precipitator with Porous Carbon Electrodes to Collect Carbon Particles". Exhaust gases from internal combustion engines contain fine carbon particles. If a biofuel is used as the engine fuel for low-carbon emission, the exhaust gas still contains numerous carbon particles. For example, the ceramic filters currently used in automobiles with diesel engines trap these carbon particles, which are then burned during the filter regeneration process, thus releasing additional CO_2. Electrostatic precipitators are generally suitable to achieve low particle concentrations and large treatment quantities. However, low-resistivity particles, such as carbon particles, cause re-entrainment phenomena in electrostatic precipitators. In this study, the author developed an electrostatic precipitator to collect fine carbon particles. Woodceramics were used for the grounded electrode in the precipitator to collect the carbon particles on the carbon electrode. Woodceramics electrodes had higher resistivity and roughness compared with those of stainless-steel electrodes. We evaluated the influence of woodceramics electrodes on the electric field formed by electrostatic precipitators and calculated the corresponding charge distribution. Furthermore, the particle-collection efficiency of the developed system was evaluated using an experimental apparatus.

Yoshida [6] contributed a paper entitled "Fundamental Evaluation of Thermal Switch Based on Ionic Wind". The author described that a significant amount of thermal energy (mainly under 200 °C) is wasted across the world. To utilize the waste heat, efficient heat management and thermal switching are required. In this paper, the basic characteristics of a thermal switch that controls the flow of heat by switching on/off the ionic wind were discussed. The study was conducted through experiments and numerical simulations. A heater made of aluminum block maintained at 100 °C was used as a heat source, and the rate of heat flow to a copper plate placed over it was measured. Ionic wind was induced by corona discharge with a needle placed on the heater. The ratio of heat transfer coefficients was obtained in the range of 3–4, with an energy efficiency of approximately 10. The heat flux at this condition was approximately 400 W/m^2. The numerical simulations indicated that the heat transfer was enhanced by ionic winds, and the results were observed to be consistent with the experimental ones. The numerical prediction of heat transfer using the ionic wind is a novel result, and future research and development can be expected.

Zukeran et al. [7] contributed a paper entitled "Collection Characteristic of Nanoparticles Emitted from a Diesel Engine with Residual Fuel Oil and Light Fuel Oil in an Electrostatic Precipitator". The purpose of the study was to investigate the collection characteristics of nanoparticles emitted from a diesel engine in an electrostatic precipitator (ESP). The experimental system consisted of a diesel engine (400 mL) and an ESP; residual fuel oil and light fuel oil were used in the engine. Although the peak value of distribution decreased as the applied voltage increased owing to the electrostatic precipitation effect, the particle concentration, at a size of approximately 20 nm, increased compared with that at 0 kV in the exhaust gas from the diesel engine with residual fuel oil. However, the efficiency was increased by optimizing the applied voltage, and the total collection efficiency in the exhaust gas, using the residual fuel oil, was 91%. In contrast, the particle concentration, for particle diameters smaller than 20 nm, did not increase in the exhaust gas from the engine with light fuel oil. Zukeran et al. are an expert group on ESPs and we believe their study will be a great success in the future.

Kuwahara et al. [8] contributed a result of high reduction efficiencies of adsorbed NO_x in pilot-scale after-treatment using nonthermal plasma in marine diesel-engine exhaust gas. The marine diesel-engine exhaust gas is one of the recent targets to be treated from the viewpoint of global environmental protection [9]. In this paper, an efficient NO_x reduction aftertreatment technology for a marine diesel engine that combines nonthermal plasma (NTP) and NO_x adsorption/desorption was reported. The aftertreatment technology can also treat particulate matter using a diesel particulate filter and regenerate it via NTP-induced ozone. The investigated marine diesel engine generates 1 MW of output power at 100% engine load. NO_x reduction was performed by repeating the adsorption/desorption processes with NO_x adsorbents and NO_x reduction using NTP. Experiments were performed for a larger number of cycles compared with those in our previous study; the amount of adsorbent used was 80 kg. The relationship between the mass of desorbed NO_x and the energy efficiency of NO_x reduction via NTP was established. This aftertreatment achieved a high reduction efficiency of 71% via NTP and a high energy efficiency of 115 $g(NO_2)/kWh$ for a discharge power of 12.0 kW. This is a significant value for marine NO_x treatment in the exhaust gas.

Energies 2019, 12, 4416

3. Conclusions

Plasma is an effective way to make, use, or treat gas. Plasma is also effective to build a better life. Furthermore, to clean the atmospheric environment, which has been polluted by fossil fuel exhaust gases, and to regain blue skies around the world, exhaust gas aftertreatments for thermal power plants and vehicles are indispensable. However, it is not necessary to replace the existing exhaust gas aftertreatment system, such as the selective catalytic reduction method, for thermal power plants and vehicles with environmental plasma technologies. From a global perspective, the majority of combustion systems do not have an exhaust gas aftertreatment system, mainly in developing countries. Plasma treatment should be an effective low-cost method for mitigating this problem. In particular, the wet NO_x treatment method via ozone injection has been attracting attention because the cost of plasma devices has recently decreased. This system should attract further attention in the future.

In addition to the exhaust gas cleaning from combustion equipment, the equipment and concepts of cleaning machines for PM, NO_x, and CO_2 that have already diffused into the atmospheric air are promising. The concepts of atmospheric air cleaners, such as "cleaning equipment for the atmosphere", which uses renewable energy sources or power generated by nuclear power plants, and "cars that can clean the air", which can use surplus power from electric vehicles, have already been proposed. The air cleaner concept is already used in various industries such as an air cleaner in a closed space of a subway station platform contaminated with PM generated by the friction of the train wheels. In addition, there are significant advances in plasma environmental cleaning technology, and there is a possibility of application to marine diesel engines [9]. We look forward to the future development of various environmental plasma technologies reported in this special issue.

Acknowledgments: The authors are grateful to MDPI for the invitation to act as guest editors for this special issue and are indebted to the editorial staff of *Energies* for their kind cooperation, patience, and committed engagement.

Conflicts of Interest: The author declares no conflict of interest.

References

1. Liu, S.; Zhao, N.; Zhang, J.; Yang, J.; Li, Z.; Zheng, H. Experimental and Numerical Investigations of Plasma Ignition Characteristics in Gas Turbine Combustors. *Energies* **2019**, *12*, 1511. [CrossRef]
2. Ahn, H.M.; Jang, E.; Ryu, S.-H.; Lim, C.S.; Lee, B.K. Control Strategy for Power Conversion Systems in Plasma Generators with High Power Quality and Efficiency Considering Entire Load Conditions. *Energies* **2019**, *12*, 1723. [CrossRef]
3. Tamošiūnas, A.; Gimžauskaitė, D.; Aikas, M.; Uscila, R.; Praspaliauskas, M.; Eimontas, J. Gasification of Waste Cooking Oil to Syngas by Thermal Arc Plasma. *Energies* **2019**, *12*, 2612. [CrossRef]
4. Yamasaki, H.; Koizumi, Y.; Kuroki, T.; Okubo, M. Plasma–Chemical Hybrid NO_x Removal in Flue Gas from Semiconductor Manufacturing Industries Using a Blade-Dielectric Barrier-Type Plasma Reactor. *Energies* **2019**, *12*, 2717. [CrossRef]
5. Kawada, Y.; Shimizu, H. Development of an Electrostatic Precipitator with Porous Carbon Electrodes to Collect Carbon Particles. *Energies* **2019**, *12*, 2805. [CrossRef]
6. Yoshida, K. Fundamental Evaluation of Thermal Switch Based on Ionic Wind. *Energies* **2019**, *12*, 2963. [CrossRef]
7. Zukeran, A.; Sawano, H.; Yasumoto, K. Collection Characteristic of Nanoparticles Emitted from a Diesel Engine with Residual Fuel Oil and Light Fuel Oil in an Electrostatic Precipitator. *Energies* **2019**, *12*, 3321. [CrossRef]
8. Kuwahara, T.; Yoshida, K.; Kuroki, T.; Hanamoto, K.; Sato, K.; Okubo, M. High Reduction Efficiencies of Adsorbed NO_x in Pilot-Scale Aftertreatment Using Nonthermal Plasma in Marine Diesel-Engine Exhaust Gas. *Energies* **2019**, *12*, 3800. [CrossRef]
9. Okubo, M.; Kuwahara, T. *New Technologies for Emission Control in Marine Diesel Engines*, 1st ed.; Butterworth-Heinemann, Elsevier: Oxford, UK, 2019; ISBN1 9780128123072. ISBN2 9780128123089.

energies

MDPI

Article

Experimental and Numerical Investigations of Plasma Ignition Characteristics in Gas Turbine Combustors

Shizheng Liu, Ningbo Zhao *, Jianguo Zhang, Jialong Yang *, Zhiming Li and Hongtao Zheng

College of Power and Energy Engineering, Harbin Engineering University, Harbin 150001, China;
liushizheng1990@163.com (S.L.); 18646297262@163.com (J.Z.); lizhimingheu@126.com (Z.L.);
zhenghongtao9000@163.com (H.Z.)
* Correspondence: zhaoningbo314@126.com or zhaoningbo314@hrbeu.edu.cn (N.Z.);
  yangjialongheu@126.com (J.Y.); Tel.: +86-0451-8251-9647 (N.Z.)

Received: 13 March 2019; Accepted: 17 April 2019; Published: 22 April 2019

Abstract: Reliable ignition is critical for improving the operating performance of modern combustor and gas turbines. As an alternative to the traditional spark discharge ignition, plasma assisted ignition has attracted more interest and been shown to be more effective in increasing ignition probability, accelerating kernel growth, and decreasing ignition delay time. In this paper, the operating characteristic of a typical self-designed plasma ignition system is investigated. Based on the optical experiment, the plasma jet flow feature during discharge is analyzed. Then, a detailed numerical study is carried out to investigate the effects of different plasma parameters on ignition enhancement of a one can-annular combustor used in gas turbines. The results show that plasma indeed has a good ability to expand the ignition limit and decrease the minimum ignition energy. For the studied plasma ignitor, the initial discharge kernel is not a sphere but a jet flow cone with a length of about 30 mm. Besides, the numerical comparisons indicate that the additions of plasma active species and the increases of initial energy, plasma jet flow length and discharge frequency can benefit the acceleration of kernel growth and flame propagation via thermal, kinetic and transport pathways. The present study may provide a suitable understanding of plasma assisted ignition in gas turbines and a meaningful reference to develop high performance ignition systems.

Keywords: plasma; ignition; gas turbine; combustor

---

1. Introduction

Ignition reliability is a key index in designing combustors because it directly affects the operation performance of gas turbines and their based power plant. In recent years, driven by the need for energy conservation and emissions reduction, many lean combustion concepts including twin annular premixing swirler (TAPS) [1], lean direct injection (LDI) [2], lean premixed prevaporized (LPP) [3], trapped vortex combustion (TVC) [4], flameless combustion (FC) [5,6], and pressure gain combustion (PGC) [7] were developed and attracted considerable attention. However, due to the fact that lean mixtures have slow flame speeds and a highly unstable flame, reliable ignition becomes one of the biggest challenges in the practice of lean combustion-based gas turbines. Besides, if affected by wet air or carbon deposition, the combustor of gas turbines is usually inevitably confronted with performance degradations of fuel spray nozzles and air swirlers, which can decrease ignition probability [8]. Therefore, developing an effective technology to achieve the reliable and robust ignition of gas turbines under various extreme operation conditions is urgently needed.

Initial kernel formation with energy deposition, early kernel growth to generate flame, flame stabilization and propagation to the whole reaction zone are the typical phases of ignition in gas turbine combustors [9,10]. In order to achieve ignition enhancement, there are three types of pathways [11–15]: thermal, kinetic and transport. Thermal enhancement is increasing the reactant temperature to accelerate the chemical reaction rate according to temperature-sensitive Arrhenius dependence. Kinetic enhancement is realized by decreasing activation energies with the addition of many active key species and radicals, which can effectively accelerate, bypass or modify the slow initiation reaction pathways. Transport enhancement is accomplished by increasing the early kernel size (greater than the so-called critical flame initiation radius) and motion with multi-channels/points or jet flow.

Plasma, considered as a distinct "fourth state of matter", provides an unprecedented opportunity for ignition control and enhancement owing to its unique capabilities in fast thermal heating via electron collision [16], producing active species (such as O, H, OH, O_3, HO_2, and NO) [17,18], reforming fuel from large molecules to small ones [19] and increasing kernel size and reactant mixing via ionic wind [20]. Over the past few decades, a large number of experimental [21–28] and numerical [29–36] investigations have been carried out to study the performance and mechanism of various plasma assisted ignition systems. Mariani et al. [37] measured the ignition performance of radio frequency sustained plasma in engines and observed that plasma had a great ability to remarkably decrease ignition temperature and extend lean limit. Wang et al. [38], Hwang et al. [39], Wolk et al. [40], Michael et al. [41], Ikeda et al. [42] and Le et al. [43] respectively experimentally investigated the ignition characteristics of microwave plasma under various operating conditions. Their results consistently showed that compared to traditional spark thermal ignition, microwave assisted plasma ignition not only significantly extended the lean limit (about 20%), but also greatly improved flame stability due to the large kernel volume and the high amount of active species. Meanwhile, they also indicated that the ignition ability of plasma was heavily dependent on the discharge type, ignitor structure and operating parameters. Sun et al. [44,45] measured the effects of nanosecond pulsed plasma on ignition and extinction of CH_4–O_2–He diffusion flames and demonstrated that the non-equilibrium plasma generated by nanosecond pulsed discharge could make the conventional S-curve with separated ignition and extinction limits degenerate to the stretched S-curve without ignition or extinction limit, which thereby enhanced ignition. Besides, other studies [46–48] revealed that owing to non-equilibrium plasma, nanosecond pulsed discharge promoted the transition from the early ignition kernel to a self-propagating flame, and the increase of pulse frequency could effectively accelerate the growth of the kernel and reduce ignition delay time and minimum ignition energy. Using the well-defined counter-flow combustion system, Ombrello et al. [49] experimentally and numerically studied the kinetic ignition enhancement of CH_4–air and H_2–air diffusion flames by non-equilibrium magnetic gliding arc plasma. It was found that plasma discharge of air leaded to significant kinetic ignition enhancement, illustrated by large decreases in ignition temperature for a broad range of strain rates. They also stated that a combination of thermal/equilibrium plasma and non-thermal/non-equilibrium plasma might be a better choice for ignition enhancement. More research investigations into plasma assisted ignition can be found in the papers by Ju and Sun [50,51] and Starikovskiy and Aleksandrov [52].

Up until now, although significant progress has been made in the validation of plasma assisted ignition, the detailed enhancement mechanisms are still not clear because of the complex multi-scale physical and chemical interactions between plasma and flame. Besides, for the present experimental investigations, most of them are performed using simple lab burners, such as constant volume combustion chambers, counter-flow systems, flow tunnels, and swirled flow reactors, etc., but limited to one focus on the gas turbine combustor under the actual operating conditions. Moreover, in terms of numerical simulation, the plasma ignition kernel is usually replaced by a spherical heat source and ignores the effects of jet flow and active species, which further restrict the understanding of the plasma ignition mechanism and the design of advanced plasma ignition systems in practice. So, it is very necessary to do more investigations on plasma assisted ignition.

In this study, an unpublished self-designed plasma ignition system which has been successfully used in a gas turbine is presented. Firstly, the discharge and jet flow characteristics of the plasma ignitor in air are optically measured to obtain the actual shape of the initial kernel. Then, taking one can-annular combustor of a gas turbine as a sample and based on the above experimental results, the ignition process is numerically analyzed. Finally, the effects of several key factors including the initial energy, concentration of active species generated by plasma, plasma jet flow length and discharge frequency on combustor ignition performance are discussed in detail.

2. Experiment Setup and Numerical Strategy

2.1. Experiment Setup

Figure 1 presents the test rig to measure the plasma jet flow characteristic during discharge in air. The basic experiment setup mainly consists of a plasma ignition system, a visualization measurement system, and a data acquisition and control system. As shown in Figure 1, the plasma ignition system is composed of the plasma ignitor, high voltage power source and cable. Once the high voltage pulse energy is delivered to the ignitor, a strong electric field between the anode and cathode will be established. When the electric field strength exceeds the breakdown threshold of air or a combustible mixture, discharge channels and the initial kernel are established. In order to obtain a large ignition kernel, several holes in the cathode wall and a unique structure for the anode are designed. More detailed information on the plasma ignition system can be found in [53]. Besides, the images of plasma jet flow during the discharge process are recorded by a high-speed camera (Phantom V7.3) with over 190×10^3 fps in the standard mode. The data acquisition and control system is used to trigger the high voltage power source and camera, record the discharge images, and change the discharge pulse parameters including voltage, frequency, and width.

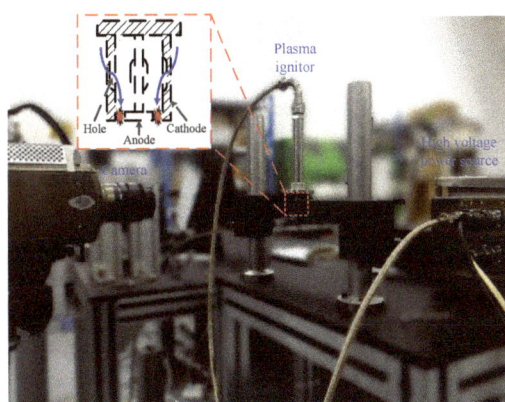

Figure 1. Schematic of the plasma ignition experiment setup.

2.2. Numerical Strategy

In the present study, a typical can-annular combustor (as shown in Figure 2) designed for gas turbines is used to numerically analyze the effects of plasma parameters on the ignition process. The length and outer diameter of the combustor are 760 mm and 1255 mm, respectively. There are 10 primary holes with a diameter of 14 mm, five dilution holes with a diameter of 13 mm (up) and 16 mm (down), and 10 rows cooling holes with diameter of 1–1.5 mm.

Comprehensively considering the basic ignition characteristics and the simulation ability of the computer, a simple physical model shown in Figure 3a is selected as the computational domain. All of its geometric parameters are consistent with the corresponding ones in Figure 2. Figure 3b presents the location of the plasma ignitor. The structured grids are generated by ANSYS ICEM to discretize the computational domain, and the grid densities near the ignition zone are sufficiently high. After the grid convergence and mesh independence validations, the final total grid number used in this numerical study is 360,000.

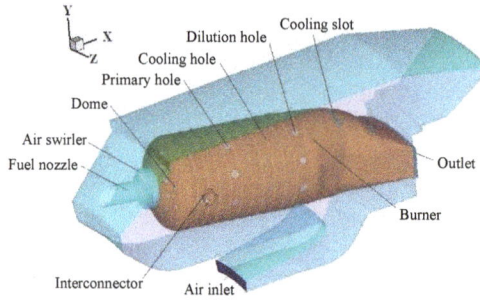

Figure 2. Can-annular combustor of a gas turbine.

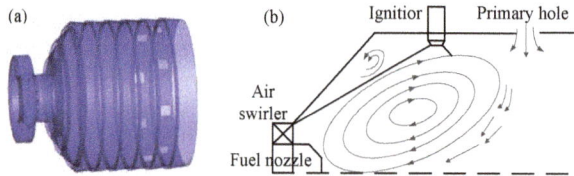

Figure 3. (**a**) Three-dimensional (3-D) computational domain and (**b**) 2-D profile.

Based on our previous numerical comparison and analysis [54], the ideal gas assumption and pressure-based Navier–Stokes solver are employed to solve the equations. The viscosity of the reactant mixture is considered, and the realizable k-ε turbulence model is selected. Gravity, buoyancy, thermophoretic force, and radiation heat transfer are ignored. The eddy dissipation concept (EDC) combustion model and cone spray model are employed. The numerical time step is 0.1 ms. The reaction rate constant is calculated using the Arrhenius formula. Due the reasonable computational cost and to discuss the effects of several plasma active key species on ignition enhancement, a reduced detail chemical mechanism for air–$C_{12}H_{23}$ (12 species and 10 steps [55]) including the intermediate species of O, OH, and CO, is used.

During numerical simulation, mass flow inlet boundaries are used for the inlet of air (0.19875 kg/s) and $C_{12}H_{23}$ (0.00600 kg/s) which are the actual values of one operating condition of a gas turbine. The temperature and pressure of air are 366 K and 0.218 MPa, respectively. The outlet selects the pressure outlet boundary. The ignition process simulation is realized by user defined function (UDF).

Figure 4 shows the numerical temperature fields of a can-annular combustor in Li [54] and the above simple model. The comparison shows that there is little difference between them. This means that it is feasible to numerically study the ignition process of a gas turbine combustor using the presented simple computational domain and numerical approach.

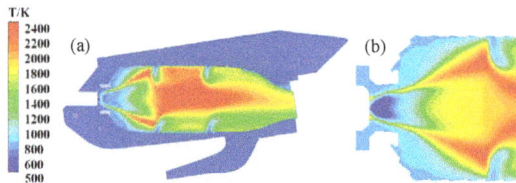

Figure 4. Numerical results of (**a**) Li [54] and (**b**) the present study.

3. Results Analysis and Discussion

3.1. Plasma Jet Flow Characteristics During Discharge

As mentioned above, the initial kernel size is an important factor affecting the ignition process and directly dependent on the geometric and discharge characteristic of the plasma ignitor. To better capture the jet flow information of plasma, the frequency and width of the discharge pulses are properly increased in the present test. Based on the experiment setup shown in Figure 1, Figure 5 images one discharge process of the self-designed plasma ignition system. As shown in Figure 5, a small initial discharge kernel is generated at time of 0.4 ms with the trigger of a high voltage pulse. Then, due to the design of the holes in cathode wall and the unique structure of the anode, the generated plasma or the heated air expand quickly, which is the so-seen jet flow. After 2.4 ms, the jet flow size begins to decrease gradually due to the interruption of the discharge.

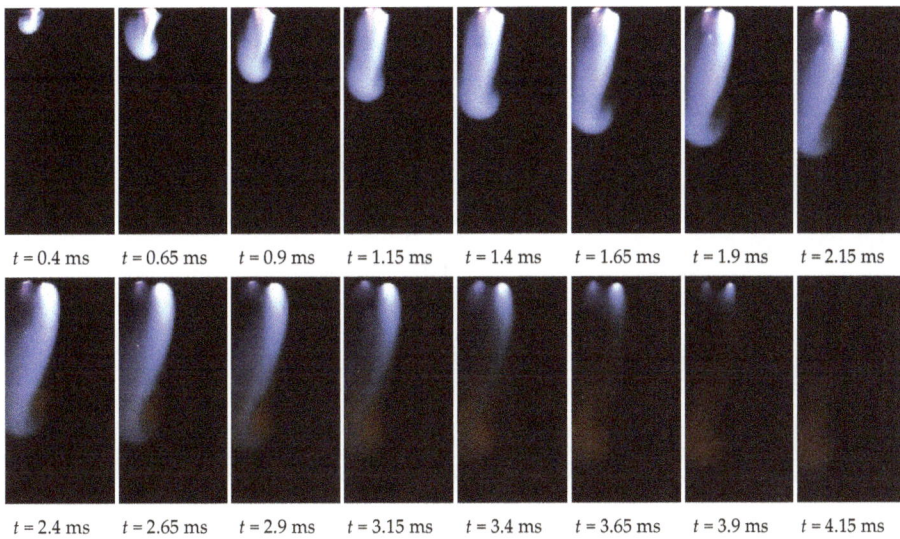

| $t = 0.4$ ms | $t = 0.65$ ms | $t = 0.9$ ms | $t = 1.15$ ms | $t = 1.4$ ms | $t = 1.65$ ms | $t = 1.9$ ms | $t = 2.15$ ms |

| $t = 2.4$ ms | $t = 2.65$ ms | $t = 2.9$ ms | $t = 3.15$ ms | $t = 3.4$ ms | $t = 3.65$ ms | $t = 3.9$ ms | $t = 4.15$ ms |

Figure 5. Plasma jet flow sequence during discharge.

Besides, a careful inspection of Figure 5 reveals that for the designed plasma ignition system, the discharge kernel is cylindrical or cone-shaped which is significantly different from the normal sphere obtained by most conventional spark or high-energy ignition systems. In addition, according to the further quantitative analysis (as shown in Figure 6), it is found that the maximum jet flow size has a length of 30 mm and a diameter of 10 mm. The kernel filled by active species and high temperatures (which is the highlighted area in Figure 6) has a length of 10 mm and a diameter of 6 mm, which is larger than those of many other spark or plasma assisted ignition systems [56,57]. This may mean that the present plasma ignition system has great potential to enhance the ignition performance of combustors and to improve the operational ability of engines.

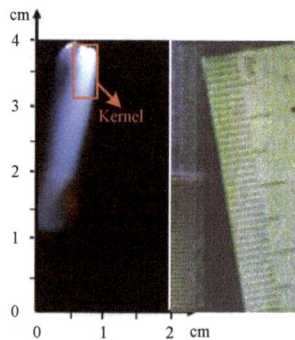

Figure 6. Plasma jet flow size at $t = 2.4$ ms.

3.2. Ignition Process Analysis in Gas Turbine Combustors

Considering the above plasma jet flow characteristic, Figures 7 and 8 respectively present the profile and cross section (X = 130 mm) numerical temperature field evolution of a successful ignition process of a can-annular combustor. Before ignition, the temperature of air is so low that the evaporation rate of the fuel droplet is very slow. Once a discharge kernel is generated, the produced high temperature can accelerate the evaporation of the droplet [10,14,16,58]. When the discharge heat energy is high enough to trigger the chemical activity of the reactant, a local ignition kernel is formed. Subsequently, owing to the comprehensive effects of plasma discharge in flow jet and active species (mainly including O, OH and CO, etc.), the initial ignition kernel gradually grows and develops into a flame, as shown at $t = 6$ ms. Then, the local flame rapidly propagates to the surrounding combustible reactant in the secondary backflow zone. The temperature field at the time of 50 ms shows that the tangential flame spread area is affected by the strong shear stress of the swirled air, and is now obviously larger than the axial one. Later, with the propagation of the flame to the primary hole, the flame propagation speed in the tangential direction will be slower than that in the axial direction. This is because for the present combustor geometry and boundary conditions, the axial velocity of the flow field is high, as shown in Figure 9. Besides, careful observation of Figures 7 and 8 reveals that up to a time of 150 ms, the flame still mainly propagates in the secondary backflow zone and has not yet been into the main backflow zone which has high turbulent kinetic energy and can easily cause the flame to extinguish. On this basis, at a time of 260 ms, most of the reactants in the secondary backflow zone are ignited, and due to the strong jet flow of air from the primary hole, the flame is rapidly transported to the main backflow zone. After that, driven by the main backflow field, the flame gradually propagates towards the inlet and then ignites the whole combustor head and forms the stable self-sustaining flame.

According to the above analysis, it can be concluded that both ignition parameters and flow field distribution are important factors that affect ignition performance of a combustor. In order to realize a successful ignition of a gas turbine under the actual operating condition, there are two necessary constraints. Firstly, high temperature, large volume and massive active species for the initial ignition process are needed to enhance the evaporation of the fuel droplet, the acceleration of the chemical reaction, and the stable growth of flame kernel. Secondly, the heat generated by the formed local flame must be higher than the cold flow, which ensures the stable propagation of the flame in the secondary and main backflow zones. Otherwise, the generated ignition kernel or flame cannot be effectively propagated and will be extinguished.

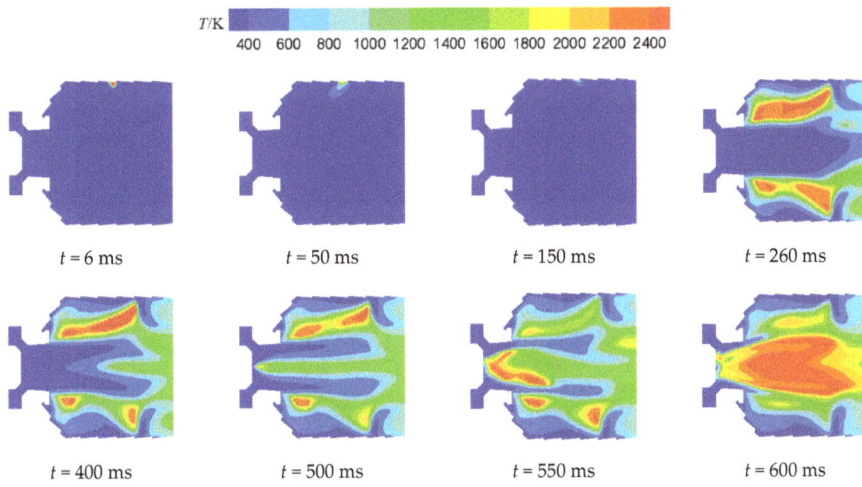

Figure 7. Numerical temperature field of combustor profile.

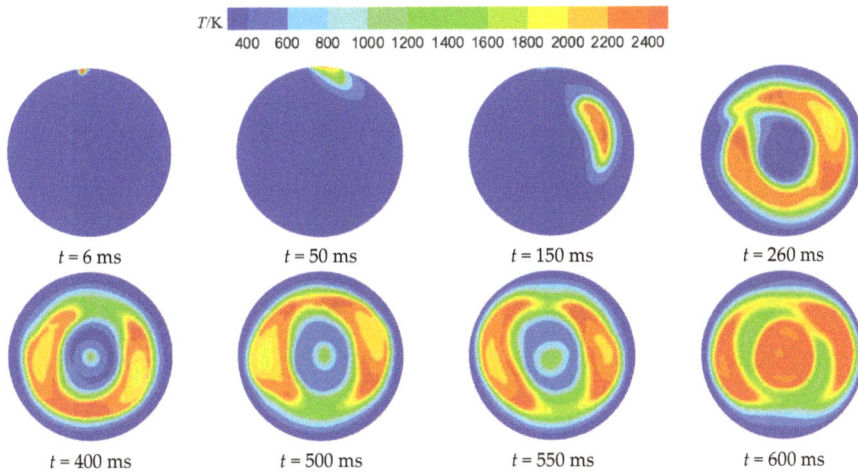

Figure 8. Numerical temperature field of combustor cross section (X = 130 mm).

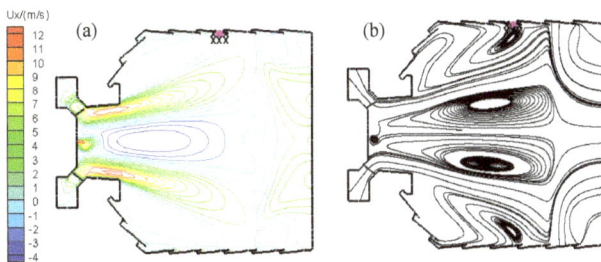

Figure 9. (a) Axial velocity and (b) streamline distribution.

3.3. Effects of Different Factors on Plasma Ignition Performance

Thermal, kinetic and transport are three well-known mechanisms for plasma ignition enhancement. In order to better understand the plasma assisted ignition process in an actual gas turbine combustor, this section will discuss the effects of the above three mechanisms by changing the initial energy, active species concentration, jet flow length and discharge frequency.

- *Energy and Active Species*

The generation, evolution and disappearance of active species during plasma discharge result from the multiscale interactions between electrons, positive and negative ions, and radicals. Such complex physical and chemical processes usually imply that the type of active species are directly dependent on the plasma discharge behavior and reactant properties. To the best of our knowledge, it is still very difficult to accurately numerically simulate the detailed characteristics of plasma active species and their effects on ignition or combustion using the commercial software ANSYS FLUENT. In order to overcome this problem, the effects of plasma active species on the air–$C_{12}H_{23}$ based ignition process is realized by simply varying the concentrations of the O atom, OH radical and CO [59,60].

Table 1 lists the detailed parameters of one-time ignition for numerical comparisons. From Table 1, it is clearly seen that under the present operating conditions, both increasing ignition energy and increasing active species can enhance the ignition performance. In the absence of plasma, the energy to ignite the combustor is not less than 400 W. However, when the active species are added, the corresponding energy can be decreased to 300 W. Besides, the comparisons of Cases A–E show that the addition of plasma active species improves the combustor ignition ability. One of the main reasons behind this is that the excited O, OH and CO will significantly contribute to the induction of the chain reaction and the enhancement of chain propagation with fuel molecules, which effectively shorten the ignition delay time. This is commonly regarded as the kinetic mechanism of plasma.

Table 1. The detailed ignition parameters and results.

Cases	Initial Kernel Radius (mm)	Initial Jet Flow Length (mm)	Ignition Energy (W)	Active Species	Results
A				0	failure
B				1% (O+OH+CO)	failure
C			300	2% O + 1% (OH+CO)	successful
D	4	12		3% O + 1.5%(OH+CO)	successful
E				4% O + 2%(OH+CO)	successful
F			400	0	failure
G			500	0	successful

Further, the results shown in Figure 10 indicate that at lower concentrations of active species, plasma can reduce the time needed for successful ignition (it is the difference between the initial time and the time when temperature is up to 660 K and chemical reaction rate is up to 2.0 kmol/m^3 s). For example, the time in Case C is about 276 ms, but the time in Case D is only 215.5 ms, which means that there was a decrease of 28%. However, in the case of strong plasma (i.e., Cases D and E), active species have little effect on ignition behavior. Compared to OH and CO, O plays a more positive role in ignition enhancement. The above results indicate that the generation and control of plasma active species is the key factor to realize the compact design and performance optimization of plasma ignition systems.

In order to further evaluate the enhancement performance of plasma ignition, Figure 11 compares the maximum excess air coefficient which can reflect the lean ignition limit under different inlet air temperatures. The greater the maximum excess air coefficient, the more lean the reactant. As pictured in Figure 11, due to the effects of active species, the lean ignition limit of the combustor can be extended. The enhancement is about 3% when the inlet air temperature varies from 280 K to 400 K. Meanwhile, the results show that with the increase of inlet air temperature, the ignition performance is also enhanced. This is because increasing air temperature is beneficial to the evaporation of fuel droplets, the mixing and transportation of the reactant, and the induction of the chemical reaction.

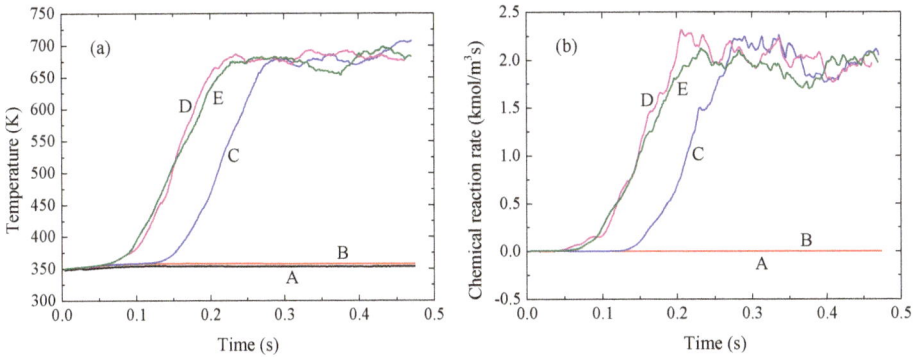

Figure 10. The changing of the combustor (**a**) average temperature and (**b**) chemical reaction rate with time in Cases A–E.

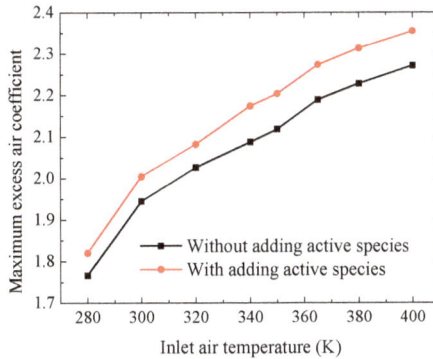

Figure 11. Effects of plasma active species on lean ignition limit.

- *Jet Flow Length of Plasma*

In practice, besides the geometry of the ignitor, the complex electromagnetic coupling phenomena can accelerate the movement of electrically charged particles, which then improve the jet flow of plasma. Motivated by this reason, this section will discuss the effects of plasma jet flow length (changing from 4 mm to 20 mm) on the ignition process of a combustor. All other parameters are consistent with those in Case C.

Table 2 gives the ignition results under different jet flow lengths. The results show that when the length is larger than 8 mm, the combustor can be ignited successfully. This effectively verifies the concept of critical flame initiation radius discussed by Chen et al. [61], Kim et al. [62], Kelley et al. [63] and Lin et al. [64]. However, in Case H, the jet flow length is so small that the high temperature and large number of active species are only gathered near the combustor wall and cannot be effectively transported into the secondary backflow zone (as shown in Figure 12a), which leads to the generated local flame kernel being very unstable. Besides, the fuel field distribution shown in Figure 12b reveals that due to the effect of velocity profile, the mass fraction of the fuel near the combustor wall is less than 0.045. Such a lean mixture will further decrease the ignition ability of a combustor.

Table 2. Combustor ignition ability under different plasma jet flow lengths.

Cases	Initial Jet Flow Length (mm)	Results
H	4	failure
I	8	successful
J	12	successful
K	16	successful
L	20	successful

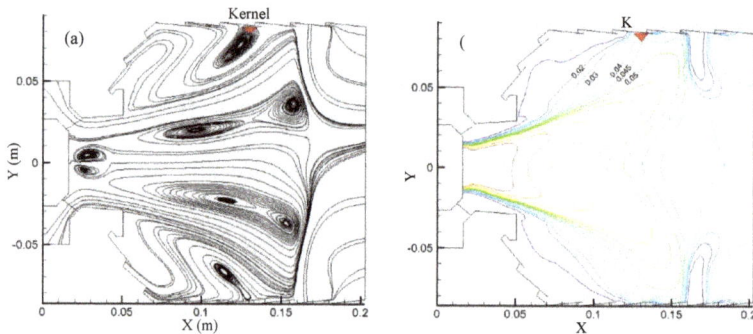

Figure 12. (**a**) Streamlines and (**b**) fuel mass fraction distributions.

On this basis, Figure 13 compares the time varying average temperature and chemical reaction rate of the combustor with different plasma jet flow lengths. The results show that with the increase of jet flow length, although the combustor can be ignited successfully, the time for successful ignition will be increased firstly and then decreased. For Cases I–L, the corresponding times for successful ignition are 211 ms, 276 ms, 251 ms and 202 ms, respectively. This is because although increasing jet flow length increases the kernel volume, the energy density is decreased as the initial ignition energy is consistent. Lower energy densities cause flame instability. So, the time for successful ignition will decrease when the jet flow length is slightly larger than the critical value (about 12 mm in this study). Then, with the further increase of jet flow length, a larger backflow zone center area and larger high mass fraction fuel area can be covered by the local flame kernel. When the positive effects exceed the adverse ones, the time for successful ignition will be shortened.

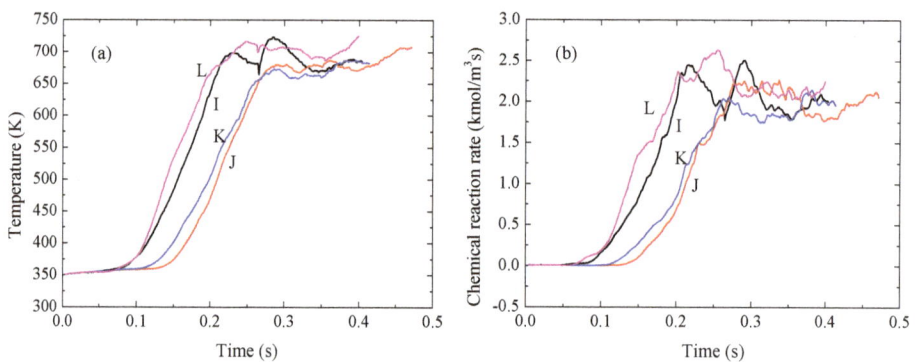

Figure 13. The changes of a combustor (**a**) average temperature and (**b**) chemical reaction rate with time in Case I–L.

From the above analysis, it can be concluded that plasma jet flow length is indeed an important factor affecting the ignition performance of a combustor. Meanwhile, both the flow and fuel fields should be considered comprehensively during the actual design of plasma ignition systems and combustors. When the ignition energy is more concentrated in the center of the backflow zone and is closer to the appropriate equivalent ratios, the ignition performance of the combustor is better.

- *Discharge Frequency*

Many available experimental investigations emphasized that at a constant discharge energy, increasing the pulse number would benefit the plasma properties and extend the lean ignition limit. In this section, the discharge frequency is varied from 19.2 Hz to 31.3 Hz, and all other parameters are consistent with those in Case B.

Table 3 lists the ignition results under different discharge frequencies. From Table 3, it is observed that the increase in discharge frequency can indeed enhance the ignition ability of a combustor. Combined with the numerical temperature fields shown in Figures 14 and 15, we found that multi-time discharge is helpful to improve the generation and propagation of the flame kernel. One of the main reasons for this is that increasing the discharge frequency can effectively enhance the transport of active species and heat to the surrounding mixture. Moreover, the results shown in Figures 14 and 15 also indicate that compared to Case N, Case O has a shorter time for successful ignition.

Table 3. Combustor ignition ability under different discharge frequencies.

Cases	Discharge Frequency (Hz)	Results
M	19.2	failure
N	23.8	successful
O	31.3	successful

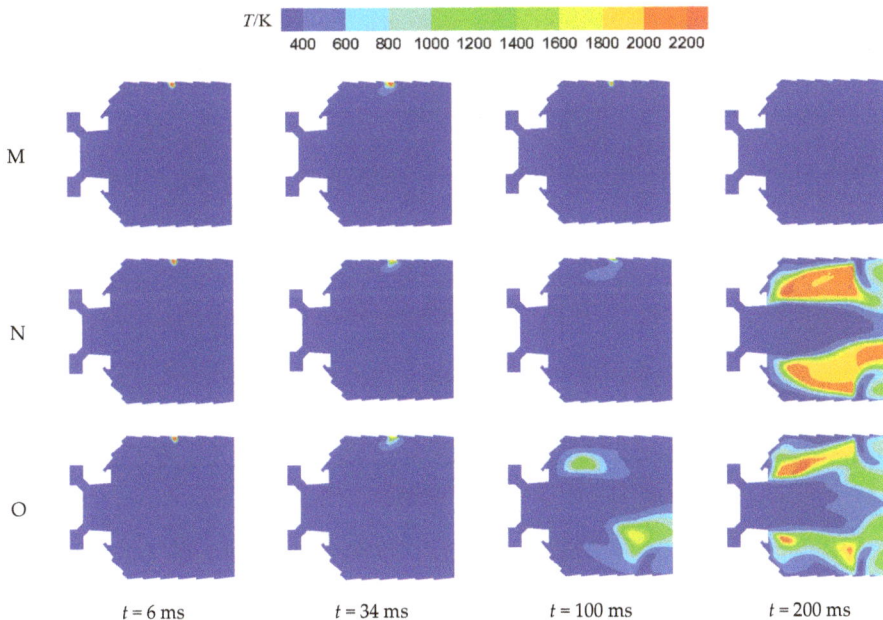

Figure 14. Numerical temperature field of combustor profile in Cases M–O.

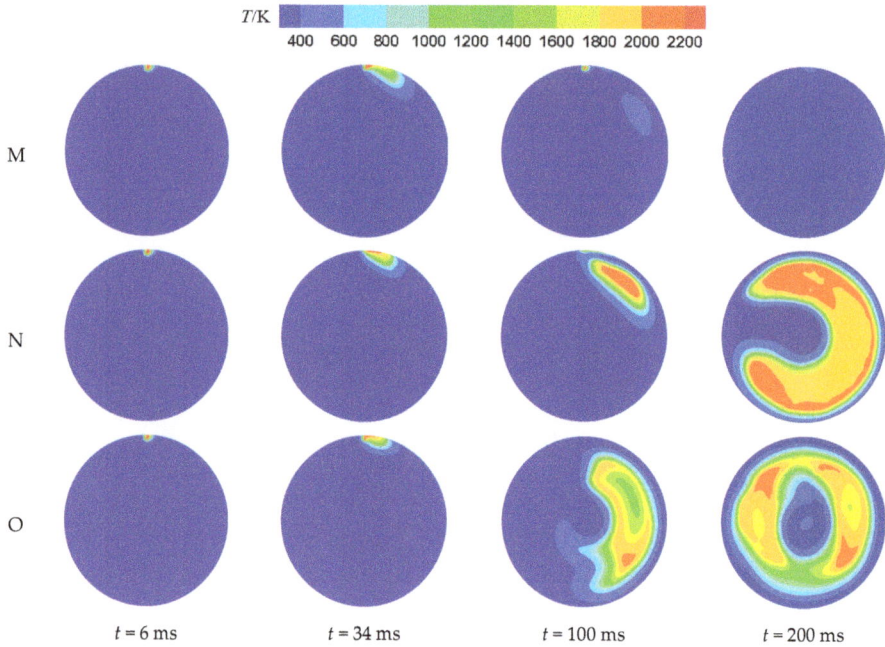

Figure 15. Numerical temperature field of combustor cross section (X = 130 mm) in Cases M–O.

To analyze the failure mechanism of Case M, Figure 16 illustrates the time varying average temperature and chemical reaction rate of the combustor. The results show that although every discharge can increase the chemical reaction rate, the formed local flame kernel is so small that it extinguishes quickly before triggering next discharge. Since the ignition energy is difficult to accumulate, the formed flame is usually very unstable and cannot self-sustain its propagation in the combustor. On the other hand, in order to realize the ignition enhancement via control discharge frequency, the time interval between two discharges should be shorter than the burnout time of the previous discharge, especial for the early phase of ignition.

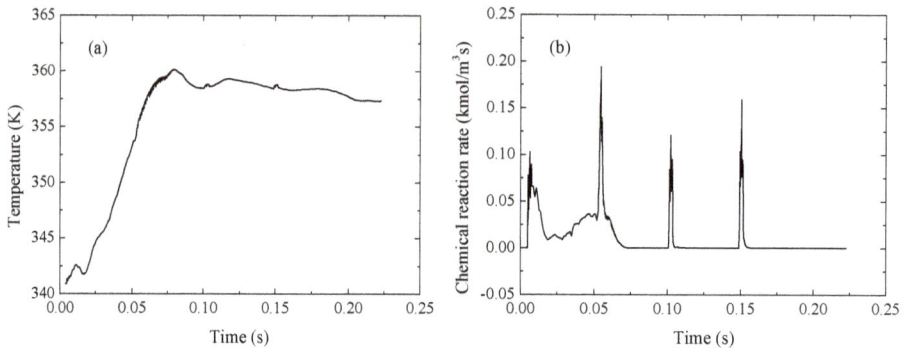

Figure 16. The changes of combustor (**a**) average temperature and (**b**) chemical reaction rate with time in Case M.

Overall, the ignition of combustor is a very complex transit chemical reaction process and its performance can be significantly affected by ignition parameters, inlet conditions and flow field distribution, etc. Therefore, in practice, the optimizing of various factors is very important to improve the ignition ability of gas turbine combustors.

4. Conclusions

The effects of plasma on the ignition process is investigated for the combustor of gas turbines. Taking one self-designed plasma ignition system as an example, high speed imaging is utilized to capture the jet flow development of plasma. On this basis, the plasma assisted ignition phase and the performance of the combustor are numerically analyzed at different ranges of initial energy, active species concentration, jet flow length and discharge frequency. The major findings of this study are summarized as follows:

(1) In contrast with the conventional spark ignitor, the present measured ignitor possesses the obvious plasma jet flow feature during discharge. Based on the effective design geometry, the jet flow length can be larger than 30 mm.

(2) The actual ignition process of a gas turbine combustor is related to the ignition parameters, backflow zone and fuel distributions. Therefore, realizing the complex optimization of different factors is the key to improve combustor ignition ability.

(3) The application of plasma can significantly enhance the ignition performance, not only for time for successful ignition but also for lean ignition limit. Besides, initial energy, active species concentration, jet flow length and discharge frequency are very critical factors affecting the ignition process. With the increase of the above four parameters, the ignition ability can be enhanced to different degrees.

(4) Although the effects of plasma on ignition is analyzed, the detailed physical and chemical process of plasma generation and evolution are not considered in this study due to limitations in the numerical approach and software. This means that many enhancement mechanisms of plasma assisted ignition cannot be clearly understood. Therefore, it is very necessary to develop an effective tool to improve the numerical precision of plasma assisted ignition.

Author Contributions: S.L. performed the numerical simulation of ignition processes under different operating conditions. N.Z. was responsible for the numerical analysis and wrote the main parts of paper. J.Z. and J.Y. did the experiments involving plasma discharge in air. Z.L. did the numerical model validation. H.Z. provided the result analysis of plasma jet flow characteristics.

Funding: This research was funded by the National Nature Science Foundation of China (Grant No. 51709059), the Province Nature Science Foundation of Heilongjiang (Grant No. QC2017045) and the Fundamental Research Funds for the Central Universities (Grant No. HEUCFJ170304, Grant No. HEUCFP201719, Grant No. HEUCFM180302).

Conflicts of Interest: We declare that we have no conflict of interest.

References

1. Foust, M.; Thomsen, D.; Stickles, R.; Cooper, C.; Dodds, W. Development of the GE Aviation Low Emissions TAPS Combustor for next Generation Aircraft Engines. In Proceedings of the 50th AIAA Aerospace Sciences Meeting Including the New Horizons Forum and Aerospace Exposition, Nashville, TN, USA, 09–12 January 2012.

2. Deepika, V.; Chakravarthy, S.R.; Muruganandam, T.M.; Bharathi, N.R. Multi-Swirl Lean Direct Injection Burner for Enhanced Combustion Stability and Low Pollutant Emissions. In Proceedings of the ASME 2017 Gas Turbine India Conference, Bangalore, India, 7–8 December 2017.

3. Gokulakrishnan, P.; Ramotowski, M.J.; Gaines, G.; Fuller, C.; Joklik, R.; Eskin, L.D.; Klassen, M.S.; Roby, R.J. A Novel Low Nox Lean, Premixed, and Prevaporized Combustion System for Liquid Fuels. *J. Eng. Gas Turbines Power* **2008**, *130*, 051501. [CrossRef]

4. Zhao, D.; Gutmark, E.; Goey, P.D. A review of cavity-based trapped vortex, ultra-compact, high-g, inter-turbine combustors. *Prog. Energy Combust.* **2018**, *66*, 42–82. [CrossRef]

5. Perpignan, A.A.V.; Rao, A.G.; Roekaerts, D.J.E.M. Flameless combustion and its potential towards gas turbines. *Prog. Energy Combust.* **2018**, *69*, 28–62. [CrossRef]

6. Khidr, K.I.; Eldrainy, Y.A.; EL-Kassaby, M.M. Towards lower gas turbine emissions: Flameless distributed combustion. *Renew. Sust. Energy Rev.* **2017**, *67*, 1237–1266. [CrossRef]
7. Snyder, P.H.; Nalim, M. Pressure Gain Combustion Application to Marine and Industrial Gas Turbines. In Proceedings of the ASME Turbo Expo 2012: Turbine Technical Conference and Exposition, Copenhagen, Denmark, 11–15 June 2012.
8. Zhao, N.B.; Wen, X.Y.; Li, S.Y. A Review on Gas Turbine Anomaly Detection for Implementing Health Management. In Proceedings of the ASME Turbo Expo 2016: Turbomachinery Technical Conference and Exposition, Seoul, Korea, 13–17 June 2016.
9. Jaravel, T.; Labahn, J.; Sforzo, B.; Seitzman, J.; Ihme, M. Numerical study of the ignition behavior of a post-discharge kernel in a turbulent stratified crossflow. *Proc. Combust. Inst.* **2019**, *37*, 5065–5072. [CrossRef]
10. Jones, W.P.; Tyliszczak, A. Large eddy simulation of spark ignition in a gas turbine combustor. *Flow Turbul. Combust.* **2010**, *85*, 711–734. [CrossRef]
11. Liao, Y.H.; Sun, M.C.; Lai, R.Y. Application of Plasma Discharges to the Ignition of a Jet Diffusion Flame. In Proceedings of the ASME 2017 Fluids Engineering Division Summer Meeting, Waikoloa, HI, USA, 30 July–3 August 2017.
12. Qiao, Y.; Mao, R.; Lin, Y. Large Eddy Simulation of the Ignition Performance in a Lean Burn Combustor. In Proceedings of the ASME Turbo Expo 2015: Turbine Technical Conference and Exposition, Montreal, QC, Canada, 15–19 June 2015.
13. Mastorakos, E. Forced ignition of turbulent spray flames. *Proc. Combust. Inst.* **2017**, *36*, 2367–2383. [CrossRef]
14. Boileau, M.; Staffelbach, G.; Cuenot, B.; Poinsot, T.; Bérat, C. LES of an ignition sequence in a gas turbine engine. *Combust. Flame* **2008**, *154*, 2–22. [CrossRef]
15. Moin, P.; Apte, S.V. Large-eddy simulation of realistic gas turbine combustors. *AIAA J.* **2006**, *44*, 698–708. [CrossRef]
16. Adamovich, I.V.; Choi, I.; Jiang, N.; Kim, J.H.; Keshav, S.; Lempert, W.R.; Samimy MUddi, M. Plasma assisted ignition and high-speed flow control: Non-Thermal and thermal effects. *Plasma Sources Sci. T.* **2009**, *18*, 034018. [CrossRef]
17. Takita, K.; Abe, N.; Masuya, G.; Ju, Y.G. Ignition enhancement by addition of NO and NO_2 from a N_2/O_2 plasma torch in a supersonic flow. *Proc. Combust. Inst.* **2007**, *31*, 2489–2496. [CrossRef]
18. Aleksandrov, N.L.; Kindysheva, S.V.; Kosarev, I.N.; Starikovskaia, S.M.; Starikovskii, A.Y. Mechanism of ignition by non-equilibrium plasma. *Proc. Combust. Inst.* **2009**, *32*, 205–212. [CrossRef]
19. Kim, W.; Mungal, M.G.; Cappelli, M.A. The role of in situ reforming in plasma enhanced ultra lean premixed methane/air flames. *Combust. Flame* **2010**, *157*, 374–383. [CrossRef]
20. Xu, K.G. Plasma sheath behavior and ionic wind effect in electric field modified flames. *Combust. Flame* **2014**, *161*, 1678–1686. [CrossRef]
21. Starikovskii, A.Y.; Anikin, N.B.; Kosarev, I.N.; Mintoussov, E.I.; Nudnova, M.M.; Rakitin, A.E.; Roupassov, D.V.; Starikovskaia, S.M.; Zhukov, V.P. Nanosecond-pulsed discharges for plasma-assisted combustion and aerodynamics. *J. Propuls. Power* **2008**, *24*, 1182–1197. [CrossRef]
22. Starikovskii, A.Y.; Anikin, N.B.; Kosarev, I.N.; Mintoussov, E.I.; Starikovskaia, S.M.; Zhukov, V.P. Plasma-assisted combustion. *Pure Appl. Chem.* **2006**, *78*, 1265–1298. [CrossRef]
23. Matveev, I.; Matveeva, S.; Gutsol, A. Non-Equilibrium Plasma Igniters and Pilots for Aerospace Application. In Proceedings of the 43rd AIAA Aerospace Sciences Meeting and Exhibit, Reno, NV, USA, 10–13 January 2005.
24. Lou, G.; Bao, A.; Nishihara, M.; Keshav, S.; Utkin, Y.G.; Rich, J.W.; Lempert, W.R.; Adamovich, I.V. Ignition of premixed hydrocarbon-air flows by repetitively pulsed, nanosecond pulse duration plasma. *Proc. Combust. Inst.* **2007**, *31*, 3327–3334. [CrossRef]
25. Guan, Y.; Zhao, G.; Xiao, X. Design and experiments of plasma jet igniter for aeroengine. *Propuls. Power Res.* **2013**, *2*, 188–193. [CrossRef]
26. Kosarev, I.N.; Kindysheva, S.V.; Momot, R.M.; Plastinin, E.A.; Aleksandrov, N.L.; Starikovskiy, A.Y. Comparative study of nonequilibrium plasma generation and plasma-assisted ignition for C_2-hydrocarbons. *Combust. Flame* **2016**, *165*, 259–271. [CrossRef]
27. Zhao, F.; Li, S.; Ren, Y.; Yao, Q.; Yuan, Y. Investigation of mechanisms in plasma-assisted ignition of dispersed coal particle streams. *Fuel* **2016**, *186*, 518–524. [CrossRef]

28. Singleton, D.; Pendleton, S.J.; Gundersen, M.A. The role of non-thermal transient plasma for enhanced flame ignition in C_2H_4-air. *J. Phys. D Appl. Phys.* **2010**, *44*, 022001. [CrossRef]
29. Mao, X.; Li, G.; Chen, Q.; Zhao, Y. Kinetic effects of nanosecond discharge on ignition delay time. *Chin. J. Chem. Eng.* **2016**, *24*, 1719–1727. [CrossRef]
30. Mao, X.; Rousso, A.; Chen, Q.; Ju, Y.G. Numerical modeling of ignition enhancement of CH_4/O_2/He mixtures using a hybrid repetitive nanosecond and DC discharge. *Proc. Combust. Inst.* **2019**, *37*, 5545–5552. [CrossRef]
31. Castela, M.; Fiorina, B.; Coussement, A.; Gicquel, O.; Darabiha, N.; Laux, C.O. Modelling the impact of non-equilibrium discharges on reactive mixtures for simulations of plasma-assisted ignition in turbulent flows. *Combust. Flame* **2016**, *166*, 133–147. [CrossRef]
32. Castela, M.; Stepanyan, S.; Fiorina, B.; Coussement, A.; Gicquel, O.; Darabiha, N.; Laux, C.O. A 3-D DNS and experimental study of the effect of the recirculating flow pattern inside a reactive kernel produced by nanosecond plasma discharges in a methane-air mixture. *Proc. Combust. Inst.* **2017**, *36*, 4095–4103. [CrossRef]
33. Casey, T.A.; Han, J.; Belhi, M.; Arias, P.G.; Bisetti, F.; Im, H.G.; Chen, J.Y. Simulations of planar non-thermal plasma assisted ignition at atmospheric pressure. *Proc. Combust. Inst.* **2017**, *36*, 4155–4163. [CrossRef]
34. Yang, S.; Nagaraja, S.; Sun, W.; Yang, V. A Detailed Comparison of Thermal and Nanosecond Plasma Assisted Ignition of Hydrogen-Air Mixtures. In Proceedings of the 53rd AIAA Aerospace Sciences Meeting, Kissimmee, FL, USA, 5–9 January 2015.
35. Yang, S.; Nagaraja, S.; Sun, W.; Yang, V. Multiscale modeling and general theory of non-equilibrium plasma-assisted ignition and combustion. *J. Phys. D Appl. Phys.* **2017**, *50*, 433001. [CrossRef]
36. Han, J.; Yamashita, H. Numerical study of the effects of non-equilibrium plasma on the ignition delay of a methane-air mixture using detailed ion chemical kinetics. *Combust. Flame* **2014**, *161*, 2064–2072. [CrossRef]
37. Mariani, A.; Foucher, F. Radio frequency spark plug: An ignition system for modern internal combustion engines. *Appl. Energy* **2014**, *122*, 151–161. [CrossRef]
38. Wang, Z.; Huang, J.; Wang, Q.; Hou, L.; Zhang, G. Experimental study of microwave resonance plasma ignition of methane-air mixture in a constant volume cylinder. *Combust. Flame* **2015**, *162*, 2561–2568. [CrossRef]
39. Hwang, J.; Bae, C.; Park, J.; Choe, W.; Cha, J.; Woo, S. Microwave-assisted plasma ignition in a constant volume combustion chamber. *Combust. Flame* **2016**, *167*, 86–96. [CrossRef]
40. Wolk, B.; DeFilippo, A.; Chen, J.Y.; Dibble, R.; Nishiyama, A.; Ikeda, Y. Enhancement of flame development by microwave-assisted spark ignition in constant volume combustion chamber. *Combust. Flame* **2013**, *160*, 1225–1234. [CrossRef]
41. Michael, J.B.; Chng, T.L.; Miles, R.B. Sustained propagation of ultra-lean methane/air flames with pulsed microwave energy deposition. *Combust. Flame* **2013**, *160*, 796–807. [CrossRef]
42. Ikeda, Y.; Nishiyama, A.; Wachi, Y.; Kaneko, M. *Research and Development of Microwave Plasma Combustion Engine (Part I: Concept of Plasma Combustion and Plasma Generation Technique)*; sae technical papers, no. 2009-01-1050; Sae International: Warrendale, PA, USA, 2019.
43. Le, M.K.; Nishiyama, A.; Serizawa, T.; Ikeda, Y. Applications of a multi-point microwave discharge igniter in a multi-cylinder gasoline engine. *Proc. Combust. Inst.* **2019**, *37*, 5621–5628. [CrossRef]
44. Sun, W.; Won, S.H.; Ombrello, T.; Carter, C.; Ju, Y.G. Direct ignition and S-curve transition by in situ nanosecond pulsed discharge in methane/oxygen/helium counter-flow flame. *Proc. Combust. Inst.* **2013**, *34*, 847–855. [CrossRef]
45. Sun, W.T.; Won, S.H.; Ju, Y.G. In situ plasma activated low temperature chemistry and the S-curve transition in DME/oxygen/helium mixture. *Combust. Flame* **2014**, *161*, 2054–2063. [CrossRef]
46. Lefkowitz, J.K.; Guo, P.; Ombrello, T.; Won, S.H.; Stevens, C.A.; Hoke, J.L.; Schauer, F.; Ju, Y.G. Schlieren imaging and pulsed detonation engine testing of ignition by a nanosecond repetitively pulsed discharge. *Combust. Flame* **2015**, *162*, 2496–2507. [CrossRef]
47. Bonebrake, J.M.; Blunck, D.L.; Lefkowitz, J.K.; Ombrello, T.M. The effect of nanosecond pulsed high frequency discharges on the temperature evolution of ignition kernels. *Proc. Combust. Inst.* **2019**, *37*, 5561–5568. [CrossRef]
48. Pancheshnyi, S.V.; Lacoste, D.A.; Bourdon, A.; Laux, C.O. Ignition of propane-air mixtures by a repetitively pulsed nanosecond discharge. *IEEE Trans. Plasma Sci.* **2006**, *34*, 2478–2487. [CrossRef]
49. Ombrello, T.; Ju, Y.G.; Fridman, A. Kinetic ignition enhancement of diffusion flames by nonequilibrium magnetic gliding arc plasma. *AIAA J.* **2008**, *46*, 2424–2433. [CrossRef]

50. Ju, Y.G.; Sun, W.T. Plasma assisted combustion: Dynamics and chemistry. *Prog. Energy Combust.* **2015**, *48*, 21–83. [CrossRef]
51. Ju, Y.G.; Sun, W.T. Plasma assisted combustion: Progress, challenges, and opportunities. *Combust. Flame* **2015**, *162*, 529–532. [CrossRef]
52. Starikovskiy, A.; Aleksandrov, N. Plasma-assisted ignition and combustion. *Prog. Energy Combust.* **2013**, *39*, 61–110. [CrossRef]
53. Yang, G.Z. Numerical Simulation and Structural Design of Continuous Plasma Generator. Master's Thesis, Harbin Engineering University, Harbin, China, 2011.
54. Li, Y.J. Research on Performance of Flame Ignition and Extinction of can Annular Combustor. PH.D. Thesis, Harbin Engineering University, Harbin, China, 2013.
55. Zhang, J.G. Numerical Simulation and Experimental Research on Plasma Ignition and Combustion Enhancement. Master's Thesis, Harbin Engineering University, Harbin, China, 2018.
56. Badawy, T.; Bao, X.; Xu, H. Impact of spark plug gap on flame kernel propagation and engine performance. *Appl. Energy.* **2017**, *191*, 311–327. [CrossRef]
57. Padala, S.; Nishiyama, A.; Ikeda, Y. Flame size measurements of premixed propane-air mixtures ignited by microwave-enhanced plasma. *Proc. Combust. Inst.* **2017**, *36*, 4113–4119. [CrossRef]
58. Lin, B.; Wu, Y.; Zhu, Y.; Song, F.; Bian, D. Experimental investigation of gliding arc plasma fuel injector for ignition and extinction performance improvement. *Appl. Energy* **2019**, *235*, 1017–1026. [CrossRef]
59. Brühl, S.P.; Russell, M.W.; Gómez, B.J.; Grigioni, G.M.; Feugeas, J.N.; Ricard, A. A study by emission spectroscopy of the active species in pulsed DC discharges. *J. Phys. D Appl. Phys.* **1997**, *30*, 2917. [CrossRef]
60. Rao, X.; Hemawan, K.; Wichman, I.; Carter, C.; Grotjohn, T.; Asmussen, J.; Lee, T. Combustion dynamics for energetically enhanced flames using direct microwave energy coupling. *Proc. Combust. Inst.* **2011**, *33*, 3233–3240. [CrossRef]
61. Chen, Z.; Burke, M.P.; Ju, Y.G. On the critical flame radius and minimum ignition energy for spherical flame initiation. *Proc. Combust. Inst.* **2011**, *33*, 1219–1226. [CrossRef]
62. Kim, H.H.; Won, S.H.; Santner, J.; Chen, Z.; Ju, Y.G. Measurements of the critical initiation radius and unsteady propagation of n-decane/air premixed flames. *Proc. Combust. Inst.* **2013**, *34*, 929–936. [CrossRef]
63. Kelley, A.P.; Jomaas, G.; Law, C.K. Critical radius for sustained propagation of spark-ignited spherical flames. *Combust. Flame* **2009**, *156*, 1006–1013. [CrossRef]
64. Lin, B.X.; Wu, Y.; Zhang, Z.B.; Chen, Z. Multi-channel nanosecond discharge plasma ignition of premixed propane/air under normal and sub-atmospheric pressures. *Combust. Flame* **2017**, *182*, 102–113. [CrossRef]

MDPI

Article

Control Strategy for Power Conversion Systems in Plasma Generators with High Power Quality and Efficiency Considering Entire Load Conditions

Hyo Min Ahn [1], Eunsu Jang [1], Seung-Hee Ryu [2], Chang Seob Lim [2] and Byoung Kuk Lee [1,*]

[1] Department of Electrical and Computer Engineering, Sungkyunkwan University, Suwon 16419, Korea; wihha@skku.edu (H.M.A.); jespro@skku.edu (E.J.)
[2] R&D Center, New Power Plasma, CO., LTD., Pyeongtaek 17703, Korea; shryu@newpower.co.kr (S.-H.R.); cslim@newpower.co.kr (C.S.L.)
* Correspondence: bkleeskku@skku.edu; Tel.: +82-31-299-4581

Received: 2 April 2019; Accepted: 2 May 2019; Published: 7 May 2019

Abstract: In this paper, a control method for the power conversion system (PCS) of plasma generators connected with a plasma chamber has been presented. The PCS generates the plasma by applying a high magnitude and high frequency voltage to the injected gasses, in the chamber. With regards to the PCS, the injected gases in the chamber could be equivalent to the resistive impedance, and the equivalent impedance had a wide variable range, according to the gas pressure, amount of injected gases and the ignition state of gases in the chamber. In other words, the PCS for plasma generators should operate over a wide load range. Therefore, a control method of the PCS for plasma generators, has been proposed, to ensure stable and efficient operation in a wide load range. In addition, the validity of the proposed control method was verified by simulation and experimental results, based on an actual plasma chamber.

Keywords: plasma generator; high-frequency DC-AC inverter; input-parallel and output-series connected inverter; two-stage AC-AC converter

1. Introduction

There has been an increasing importance for a cleaning process in a semiconductor or a display manufacturing process, as manufacturing processes have become more precise and require a high yield rate [1,2]. The purpose of the cleaning process is to remove organic contaminants and particles on the surface of a semiconductor or a display. Cleaning processes are divided into wet cleaning process (using a chemical solution) and dry cleaning (a plasma cleaning) processes. Currently, the semiconductor cleaning process is moving away from the wet cleaning process to the dry cleaning process, because of environmental issues caused by chemical wastes and the development of precise semiconductor processes [2–5]. The dry cleaning process requires a plasma generator and it comprises a power conversion system (PCS), which supplies electric power to generate plasma, and a plasma chamber that serves as a load of the PCS. In this configuration, the PCS which generates a high-frequency and high-magnitude AC voltage, is an essential part to ensure a stable and efficient operation of the plasma generator.

When controlling the PCS for the plasma generator, the characteristics of the plasma chamber should be considered. According to the condition of the chamber such as the amount of injected gases, pressure, and plasma ignition states, the required input power of the chamber to be supplied by the PCS varies. In other words, when the chamber is equivalent to the load of the PCS, the load level of the PCS is changed, depending on the condition of the chamber. The equivalent impedance of the chamber is very large (ideally open), before the ignition of the injected gases and the impedance decreases, sharply after the ignition state. Therefore, the primary requirements of the PCS for a plasma generator is that it should have a constant characteristic, to prevent a short circuit problem after ignition. Additionally, the PCS should be capable of supplying high magnitude and high frequency voltage to the chamber in the initial state, to ignite the injected gases. The next requirement is that the PCS should supply high quality power to the chamber, regardless of the chamber conditions because high quality plasma gases can be generated when the input power quality of the chamber is high. This means that the output power quality of PCS determines the quality of the plasma gases. In addition, losses of the plasma generator occur in the PCS. Thus, the efficiency of the PCS is also important for an efficient operation of the plasma generator [5–9]. In previous studies on modular pulse generators, characteristics of plasma chambers, the system efficiency, and the output power quality were not seriously considered. Therefore, they were difficult to directly apply to plasma generators requiring a stable and efficient operation [10–15].

In the previous studies on PCS for plasma generators, the characteristics of the plasma chamber were considered. However, the PCS mainly consisted of a single-module PCS, and there were several disadvantages to this configuration [6–9]. The main disadvantage arose from a transformer with a high step-up ratio or a high magnitude of input voltage for the DC-AC inverter. In this case, the losses of inverters were increased because of the increasing input current or input voltage of the inverters. In addition, the design flexibility was decreased because the hardware design should have changed according to the rated power of the plasma generator. For these reasons, in this paper, input-parallel and output-series connected inverters were applied to the PCS for the plasma generator, as shown in Figure 1. As several inverters were connected, a flexible design was possible, according to the rated power. In addition, the input voltage of a filter network ($v_{TR.S}$) was the sum of the secondary side voltages of the transformers, so a high output voltage could be generated, which was supplied to the chamber [16–19]. Another advantage was that, this configuration was not significantly affected by the unbalance between DC-AC inverters. When the secondary sides of the transformers were connected in parallel, problems of circulating currents could occur due to the unbalance between the connected DC-AC inverters. However, there was no circulating current in this configuration because the secondary sides were connected in series [18].

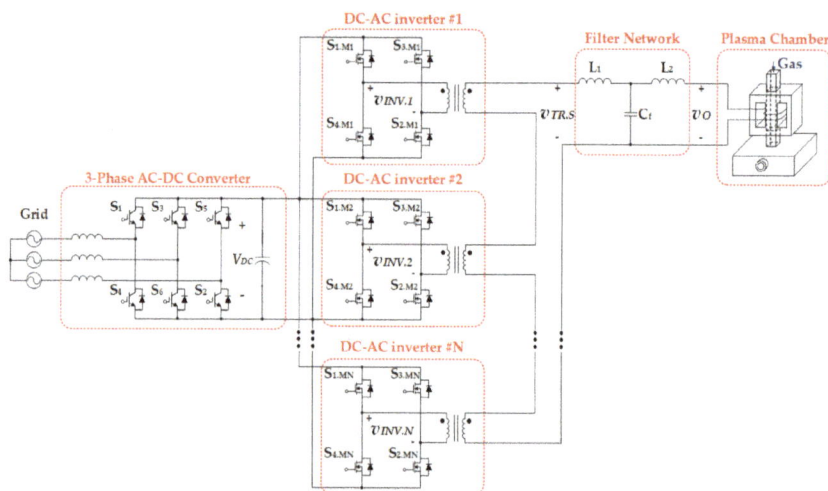

Figure 1. Configuration of the two-stage AC-AC converter for plasma generators.

Several control methods have been proposed for the two-stage AC-AC converters in Figure 1. The first method employs an AC-DC converter that controls the DC-link voltage (V_{DC}) at a constant value, and all of the connected inverters control the output power of the PCS, simultaneously [16,17]. When applying this control method, unnecessary power losses in the light and intermediate load region occur, because all of the connected inverters are used in the operation, even if one or two inverters are capable of controlling the output power. In addition, there is a disadvantage that the output power quality is decreased because THD of $v_{tr.s}$, which is the input voltage of the filter network in Figure 1, is increased. The second method employs DC-AC inverters which operate with a fixed duty ratio, and the AC-DC converter controls the V_{DC} at variable values, to control the output power [19]. When adopting this method, the quality of the output power does not degrade sharply. However, this method is not suitable for plasma generators because the impedance range of the plasma chamber is wide. Another control method employs DC-AC inverters, which control the shape of $v_{TR.S}$ as in the multi-level voltage waveforms [10,18]. However, these control methods mainly focus on the output power quality, and system efficiency is not seriously considered.

Therefore, in this paper, a control method of the two-stage AC-AC converter with input–parallel and output–series connected inverters, for plasma generators, is proposed. The proposed method controls the number of operating inverters by considering the output power level and the output power quality. Thus, by applying the proposed control method, not only the system efficiency but also the output power quality could be improved. Additionally, the validity of the proposed control method was verified through a simulation and an experiment based on an actual prototype of a PCS for plasma generators.

2. Conventional Control Algorithm of DC-AC Inverters for Plasma Generators

As shown in Figure 1, the two-stage AC-AC converter for the plasma generator consists of a three-phase AC-DC converter and DC-AC inverters. The three-phase AC-DC converter controls the input currents to ensure a high input power factor and controls the V_{DC}. The DC-AC inverters are connected in input–parallel and output–series configuration, by transformers, and each inverter performs phase-shift control with a high switching frequency, such as 400 kHz. Among these PCSs, DC-AC inverters which are directly connected to the plasma chamber mainly affect the performance of the plasma generator. The requirements of the DC-AC inverter for plasma generator are as follows. (1) As inverters operate at a high switching frequency, a zero voltage switching (ZVS) operation is necessary. (2) After the ignition of injected gases, the equivalent impedance of the chamber decreases sharply. Therefore, the filter network can operate with a constant current output, to prevent overcurrent, even though the chamber impedance drops rapidly. (3) The DC-AC inverter should be able to supply high-quality output power to a plasma chamber, for high yield rates, by generating a high-quality plasma. To satisfy the first and second requirements, selection and design of the filter network is important. Among conventional filter networks, an LCL filter network satisfies these requirements [5,20,21]. To meet the third requirements, the control method of the DC-AC inverter is important, and the conventional control method has a limitation that the output power quality decreases in the light and intermediate load region. Figure 2 shows waveforms of the DC-AC inverters that are applied in the conventional control method. All of the connected inverters simultaneously transfer power from the primary side to the secondary side of the transformers. Therefore, there are several disadvantages of the conventional control method. (1) In the light and intermediate load region, the phase shift angle (α) is the phase angle of the operating inverter switches, S_1 and S_3 in Figure 1. As shown in Equation (1), THD of $v_{TR.S}$ is high when α is small. As a result, the power quality of the output power in the light load region decreases because $v_{TR.S}$ is the input voltage of the LCL filter network. (2) The efficiency of DC-AC inverters is drastically reduced because all of the connected inverters operate, even in the load region. In other words, unnecessary switching losses occur in the light and intermediate load region, when the conventional control method is applied.

$$THD_v = \sqrt{\frac{\pi\alpha}{8\sin^2\left(\frac{\alpha}{2}\right)} - 1} \qquad (1)$$

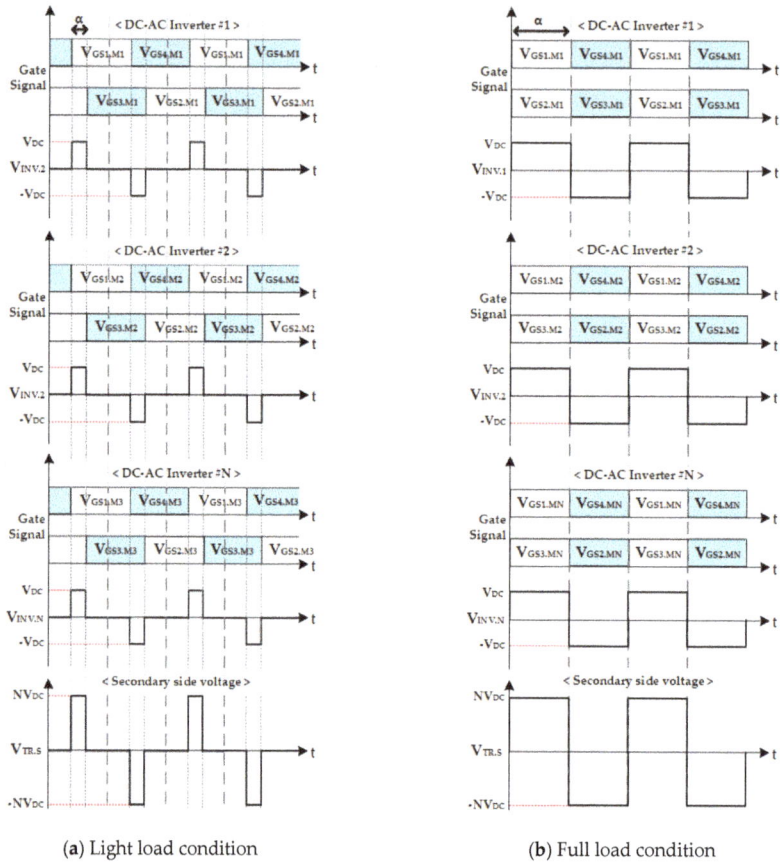

(a) Light load condition   (b) Full load condition

Figure 2. Example waveforms of DC-AC inverters with the conventional control method.

3. Proposed Control Algorithm of DC-AC Inverters for Plasma Generators

3.1. Principle and Characteristics of the Proposed Control Algorithm

As shown in Equation (1) and Figure 3, when α of the inverters become less than 0.5π, THD of $v_{TR.s}$ is increased so the quality of the output power is decreased. To solve this problem, the proposed control method controls not only α but also the number of operating inverters (N). When N is reduced, α should be increased to output the same power. Therefore, the proposed control method has a wide operating range where THD of $v_{TR.s}$ is low in the light and intermediate load region, as shown in Figure 3. By employing this control method, the degradation of output power quality can be reduced, compared with that in the conventional control method. In addition, the efficiency of the PCS can be increased because N is reduced. The proposed control method is described in detail as follows. (1) As a high output voltage is required to ignite the injected gases, all connected inverters operate with a maximum value of N and α, at the initial start. (2) After ignition, the DC-AC inverter controls α according to the reference value. (3) When α becomes less than 0.5π, N is reduced. (4) After adjusting N, α is controlled according to the reference value. (5) Steps 2 to 4 are repeated until the output power of inverters and the reference value become the same or N becomes 1. (6) If N becomes 1, the operating inverter controls α without a lower limit of α. When the reference value is smaller than the output of inverters, N is increased when α becomes π. These steps are depicted as a flowchart in Figure 4. In this flowchart, V_{REF} is the reference output voltage and V_{sen} is the sensing voltage of the output voltage of the PCS. Example waveforms for the proposed control method are shown in Figure 5. In order to prevent the injection of high-frequency induced currents to the DC-link, through the non-operating inverters, these inverters operate in a freewheeling mode (freewheeling inverters). The freewheeling current flows through the upper or lower switches. When the freewheeling current flows only through the upper or lower switches, thermal imbalance between the upper and lower switches in the same arm occurs. Therefore, as shown in Figure 6 that shows example gate signals of the operating and freewheeling inverters, the freewheeling path of freewheeling inverters is changed periodically. Additionally, the switching frequency of the freewheeling inverters should be relatively smaller than that of the operating inverters, to prevent high switching losses of freewheeling inverters.

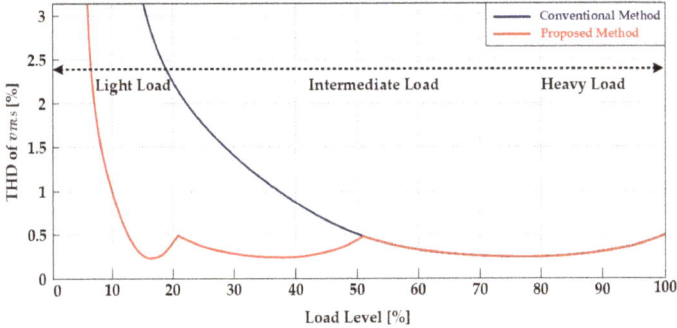

Figure 3. Comparison of operating areas according to the control methods.

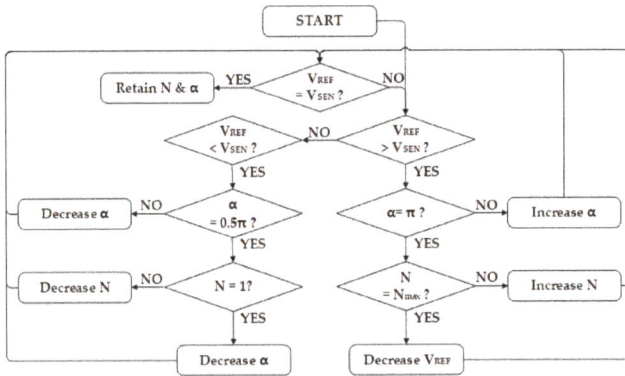

Figure 4. Flow chart of the proposed control method.

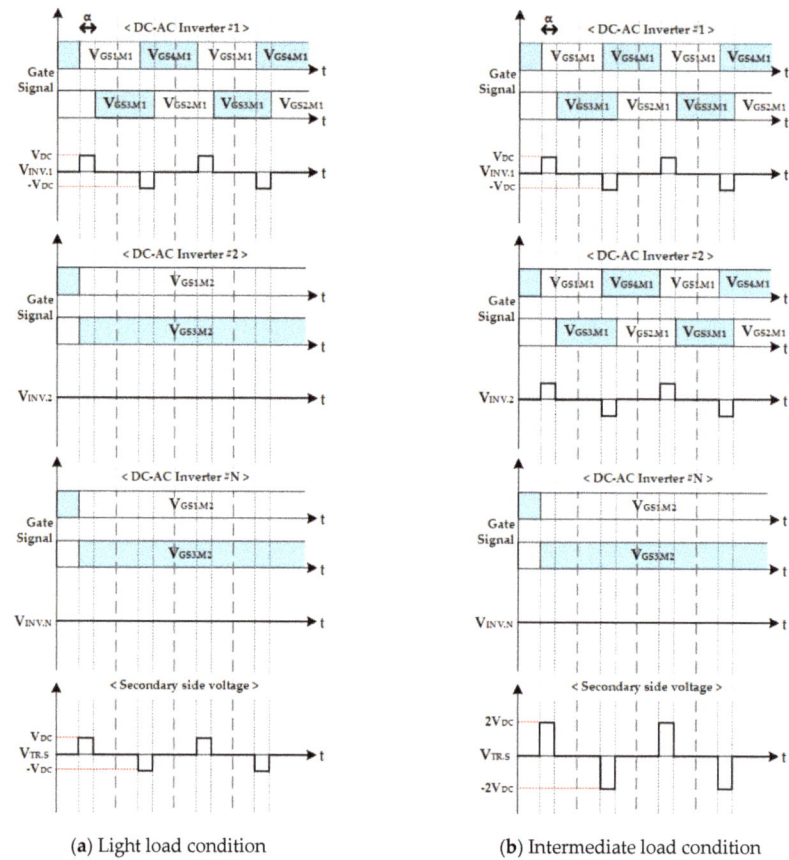

(a) Light load condition

(b) Intermediate load condition

Figure 5. Example waveforms of DC-AC inverters with the proposed control method.

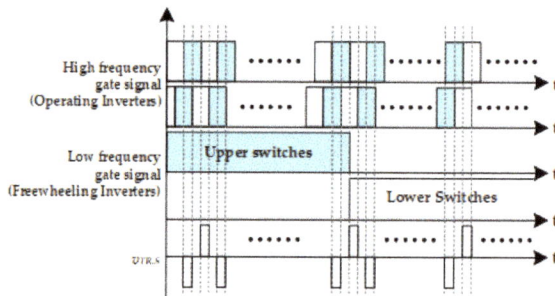

Figure 6. Gate signals of an operating inverter and a freewheeling inverter.

3.2. Mathematical Model of the Proposed Control Algorithm.

In order to apply the proposed control method and to design the corresponding hardware, it is necessary to analyze the characteristics of the input–parallel and series–connected DC-AC inverters, based on the mathematical model. The fundamental voltage and the RMS voltage of $v_{TR.S}$, in terms of N, α, and switching frequency (ω_s) are shown in Equations (2) and (3).

$$v_{TR.S(1)}(t) = \frac{4}{\pi} N V_{DC} \sin\left(\frac{\alpha}{2}\right) \sin(\omega_s t) \tag{2}$$

$$V_{TR.S(RMS)(1)} = \frac{2\sqrt{2}}{\pi} N V_{DC} \sin\left(\frac{\alpha}{2}\right) \tag{3}$$

The output voltage and the output current (i_o) can be derived from the input current (i_{INV}) and the output gain, as shown in Equations (4) and (5). In this equation, Q is the Q factor of the LCL network, γ is the ratio of the inverter-side inductor (L_1) to the load-side inductor (L_2), and ω_n is the ratio of the resonant frequency of the LCL network to the switching frequency [5,15,16].

$$H_{inv} = \frac{I_{inv}}{\left(\dfrac{V_{TR.S(RMS)(1)}}{\sqrt{L_1/C_f}}\right)} = \frac{\left(1 - \gamma\omega_n^2\right) + \frac{j\omega_n}{Q}}{\frac{1}{Q}(1 - \omega_n^2) + j[(1+\gamma)\omega_n - \gamma\omega_n^3]} \tag{4}$$

$$H_O = \frac{I_O}{\left(\dfrac{V_{TR.S(RMS)(1)}}{\sqrt{L_1/C_f}}\right)} = \frac{1}{\frac{1}{Q}(1 - \omega_n^2) + j[(1+\gamma)\omega_n - \gamma\omega_n^3]} \tag{5}$$

Using Equations (4) and (5) the output current, output voltage, and inverter-side current can be derived as Equations (6)–(8), respectively, when the equivalent chamber impedance is R_o.

$$I_{O.rms} = \frac{1}{\pi Z_n} \frac{2\sqrt{2}}{(1 - \omega_n^2)/Q + j[(1+\gamma)\omega_n - \gamma\omega_n^3]} n V_{DC} \sin\left(\frac{\alpha}{2}\right) \tag{6}$$

$$V_{O.rms} = \frac{1}{\pi Z_n} \frac{2\sqrt{2}}{(1 - \omega_n^2)/Q + j[(1+\gamma)\omega_n - \gamma\omega_n^3]} n V_{DC} R_O \sin\left(\frac{\alpha}{2}\right) \tag{7}$$

$$I_{inv.rms} = \frac{2\sqrt{2}}{\pi Z_n} \frac{\left(1 - \gamma\omega_n^2\right) + j\omega_n/Q}{(1 - \omega_n^2)/Q + j[(1+\gamma)\omega_n - \gamma\omega_n^3]} n V_{DC} \sin\left(\frac{\alpha}{2}\right) \tag{8}$$

The DC-AC inverter and magnetic components are designed based on the mathematical analysis results, and the designed hardware is simulated and experimentally studied to verify the proposed control method.

4. Simulation and Experimental Results

The design of the plasma generator described in previous sections was applied to the simulation and an experiment. The specifications of the prototype, including the three-phase AC-DC converter and DC-AC inverters, connected in input–parallel and output–series configuration, are listed in Table 1. Using Equations (2) and (3), the RMS value of $v_{TR.S}$ ($V_{TR.S(RMS)}$) was derived. To satisfy the output power condition at the calculated $V_{TR.S(RMS)}$, the required gain of the LCL filter network was 0.57 or higher. In addition, the resonant frequency of the LCL filter network for constant current operation should have been close to the switching frequency (400 kHz), so it was designed to be 450 kHz, at which the ZVS operation was possible. The parameters of the LCL filter network, which satisfied these conditions, could be derived using Equation (4), as shown in Table 2. Simulation and experiments were performed with the designed hardware, to verify the validity of the proposed control method.

Table 1. Specifications of the power conversion system (PCS) for plasma generators.

AC-DC Converter	
Parameter	**Value**
Rated Power	30 kW
Grid voltage	3Φ 440$V_{LL.RMS}$
Grid frequency	60 Hz
Switching frequency	10 kHz
DC-link voltage	800 Vdc
DC-AC Inverter	
Parameter	**Value**
Rated Power	10 kW (1 EA)
Number of connected inverters	3 EA
Switching frequency (operating inverter)	400 kHz
Switching frequency (freewheeling inverter)	10 kHz
Turn ratio of transformer	1
Equivalent resistor of chamber (R_o)	50 Ω

Table 2. Specifications of the designed LCL filter network for the PCS.

Parameter	Value
L_1	26 uH
L_2	20 uH
C_f	4.7 nF

4.1. Simulation Results

The simulation was performed using PSIM based on the system parameters in Tables 1 and 2. Three kW and 15 kW were set as the light load and intermediate load conditions, respectively. Figure 7 shows the simulation waveforms, when the proposed and conventional control methods were applied. As shown in these waveforms, α of $v_{TR.S}$ was relatively large when the proposed control method was applied, compared with α when the conventional control method was applied in the light and the intermediate load region. Therefore, the quality of the output voltage (v_{out}) was higher for the proposed control method. From the simulation results, it could be confirmed that the proposed control method was more advantageous than the conventional control method, in the light and the intermediate load region.

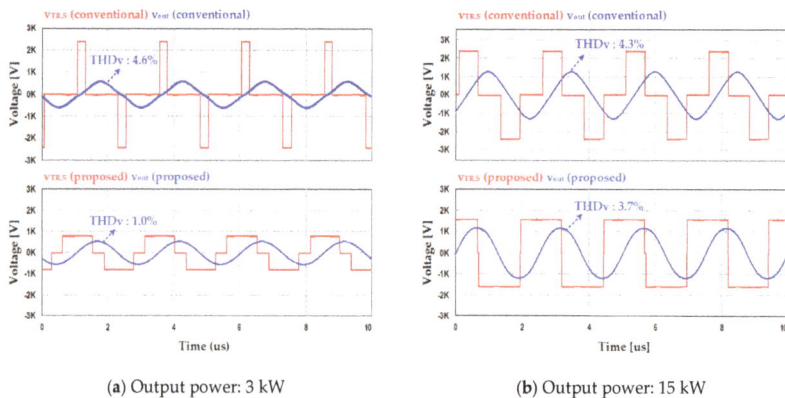

(a) Output power: 3 kW   (b) Output power: 15 kW

Figure 7. Waveforms of DC-AC inverter according to the output power and control methods.

4.2. Experimental Results

A prototype of the AC-DC converter and DC-AC inverters were designed using the same conditions as the simulation. The experimental environment is shown in Figure 8. Figure 9 shows the maximum output power, according to N and these experimental results corresponded to the simulation and the numerical analysis results. Figure 10 shows the results of a comparative experiment of the proposed and conventional control methods. When the output power exceeded 15.6 kW, the experimental waveforms of the proposed and conventional control methods were the same, because three inverters must operate in both control methods. However, when the output power was smaller than 15.6 kW, the experimental results showed that the maximum value of $v_{TR.S}$ was relatively small and α was relatively large, when applying the proposed control method. These experimental results showed that the proposed control method could improve the output power quality and increase the efficiency of the PCS, by reducing N. Finally, THDv of the output voltage and the system efficiency, according to the control methods, are presented in Figure 11. Figure 11a shows the THDv of the output voltage (i.e., the input voltage of the chamber). THDv was improved in the light load conditions when the proposed control method was applied (average 3% and maximum 3.5%). Likewise, the efficiency in the light load conditions was also improved because α of $v_{TR.S}$ was relatively large and N was small when applying the proposed control method in the light and intermediate load conditions. Therefore, the output power quality and the system efficiency were improved in the proposed control method. Both simulation and experimental results showed that the output power quality and system efficiency were improved in the light load condition, when the proposed control method was applied. Hence, these simulation results and experimental results validated the proposed control method.

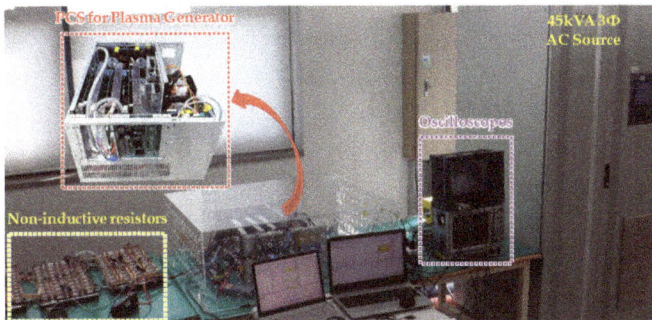

Figure 8. Experimental environment.

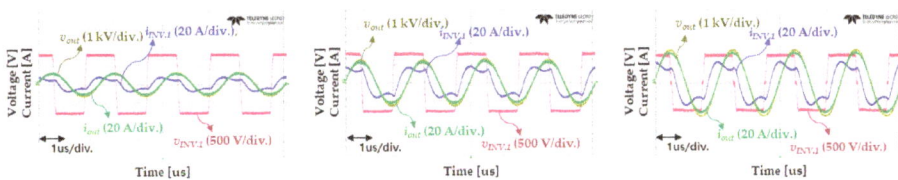

Figure 9. Waveforms in the maximum output condition according to the number of operating inverters.

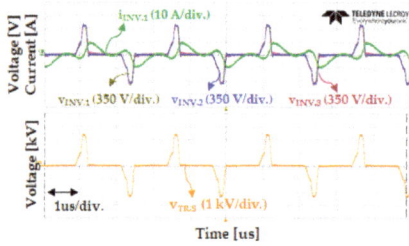

(**a**) Output power: 3 kW (Conventional method)

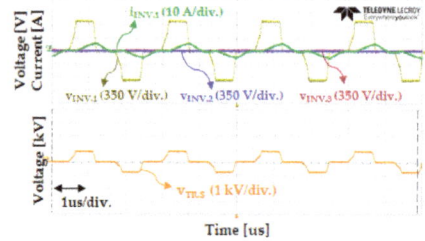

(**b**) Output power: 3 kW (Proposed method)

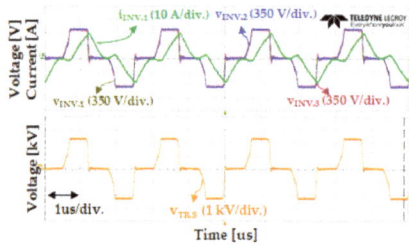

(**c**) Output power: 15 kW (Conventional method)

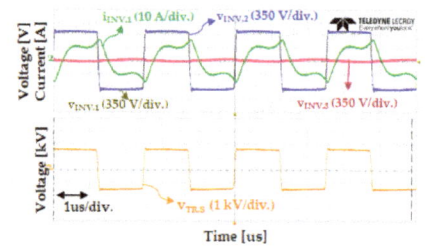

(**d**) Output power: 15 kW (Proposed method)

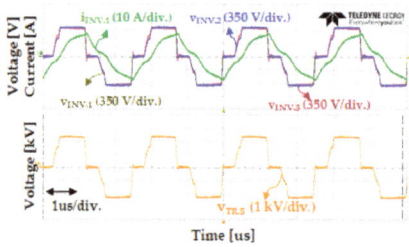

(**e**) Output power: 20 kW

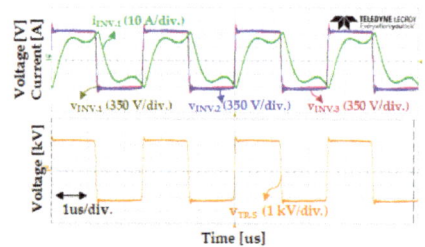

(**f**) Output power: 30 kW

Figure 10. Waveforms of inverters according to control methods.

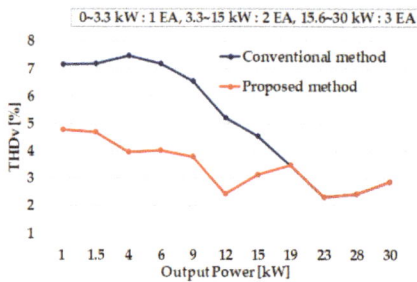

(**a**) THDv of output voltage of the PCS

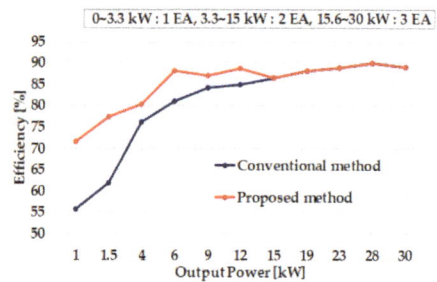

(**b**) Efficiencies of the PCS

Figure 11. Experimental results of the PCS for the plasma generator.

5. Conclusions

In this paper, a control algorithm and design for plasma generators with input–parallel and output–series connected inverters were proposed by considering the characteristics of the plasma chamber. The proposed control method controlled the number of operating inverters, according to the chamber condition and controlled the inverter operation to make it suitable for a plasma generator with an input–parallel and output–series configuration. The proposed control method and the design method were verified through informative simulation and experiment. In addition, a comparative simulation and the experiment results of the proposed control method and conventional control method were presented. The simulation and experimental results showed that the proposed control method and design method could be successfully applied to a PCS for an actual plasma generator.

Author Contributions: Conceptualization, H.M.A.; Data curation, H.M.A. and E.J.; Formal analysis, H.M.A.; Investigation, H.M.A. and E.J.; Project administration, B.K.L.; Software, H.M.A. and E.J.; Supervision, B.K.L.; Validation, H.M.A. and C.S.L.; Visualization, H.M.A. and S.-H.R.; Writing–original draft, H.M.A. and B.K.L.; Writing–review & editing, H.M.A. and B.K.L.

Funding: This work was supported by "Human Resources Program in Energy Technology" of the Korea Institute of Energy Technology Evaluation and Planning (KETEP), granted financial resource from the Ministry of Trade, Industry & Energy, Republic of Korea. (No. 20184030202190)

Conflicts of Interest: The authors declare no conflicts of interest.

References

1. Kamieniecki, E.; Foggiato, G. Analysis and control of electrically active contaminants by surface charge analysis. In *Handbook of Semiconductor Wafer Cleaning Technology*, 3rd ed.; Noyes Publications: Park Ridge, NJ, USA, 1993; pp. 497–536. ISBN 978-032-351-084-4.
2. Chen, X.; Holber, W.; Loomis, P.; Sevillano, E.; Shao, S.Q. Advances in Remote Plasma Sources for Cleaning 300 mm and Flat Panel CVD Systems. *Semiconductor® Magazine*, 20 August 2003.
3. Jin, Y.; Ren, C.S.; Fan, Q.Q.; Yan, H.; Li, Z.; Zhang, J.; Wang, D. Surface Cleaning Using an Atmospheric-Pressure Plasma Jet in O_2/Ar Mixtures. *IEEE Trans. Plasma Sci.* **2003**, *40*, 2706–2710. [CrossRef]
4. Balish, K.E.; Nowak, T.; Tanaka, T.; Beals, M. Dilute Remote Plasma Clean. U.S. Patent 6329297B1, 11 December 2001.
5. Yong-Nong, C.; Chih-Ming, K. Design of Plasma Generator Driven by High-frequency High-Voltage Power Supply. *J. Appl. Res. Technol.* **2013**, *11*, 225–234. [CrossRef]
6. Tran, K.; Millner, A. A New Power Supply to Ignite and Sustain Plasma in a Reactive Gas Generator. In Proceedings of the 2008 Twenty-Third Annual IEEE Applied Power Electronics Conference and Exposition (2008 APEC), Austin, TX, USA, 24–28 February 2008; pp. 1885–1892.
7. Kim, Y.M.; Kim, J.Y.; Mun, S.P.; Lee, H.W.; Kwon, S.K.; Suh, K.Y. The design of inverter power system for plasma generator. In Proceedings of the 2005 International Conference on Electrical Machines and Systems, Nanjing, China, 27–29 September 2005; pp. 1309–1312.
8. Ahn, H.M.; Sung, W.Y.; Lee, B.K. Analysis and Design of Resonant Inverter for Reactive Gas Generator Considering Characteristics of Plasma Load. *J. Electr. Eng. Technol.* **2018**, *13*, 345–351. [CrossRef]
9. Chen, W.; Zhuang, K.; Ruan, X. A Input-Series- and Output-Parallel-Connected Inverter System for High-Input-Voltage Applications. *IEEE Trans. Power Electron.* **2009**, *24*, 2127–2137. [CrossRef]
10. Abdelsalam, I.; Elgenedy, M.A.; Ahmed, S.; Williams, B.W. Full-bridge Modular Multilevel Submodule-based High-voltage Bipolar Pulse Generator with Low-voltage DC, Input for Pulsed Electric Field Applications. *IEEE Trans. Plasma Sci.* **2017**, *45*, 2857–2864. [CrossRef]
11. Luc-André, G.; Bélanger, J. A Modular Multilevel Converter Pulse Generation and Capacitor Voltage Balance Method Optimized for FPGA Implementation. *IEEE. Trans. Ind. Electron.* **2015**, *62*, 2859–2867. [CrossRef]
12. Elgenedy, A.M.; Darwish, A.; Ahmed, S.; Williams, B.W. A Modular Multilevel-Based High-Voltage Pulse Generator for Water Disinfection Applications. *IEEE Trans. Plasma Sci.* **2016**, *44*, 2893–2900. [CrossRef]
13. Elserougi, A.A.; Massoud, A.M.; Ahmed, S. A Modular High-Voltage Pulse-Generator with Sequential Charging for Water Treatment Applications. *IEEE. Trans. Ind. Electron.* **2016**, *63*, 7898–7907. [CrossRef]

14. Mohamed, A.; Darwish, A.; Williams, B.W. A Transition Arm Modular Multilevel Universal Pulse-Waveform Generator for Electroporation Applications. *IEEE. Trans. Power. Electron.* **2017**, *32*, 8979–8981. [CrossRef]

15. Rocha, L.L.; Silva, J.F.; Redondo, L.M. Multilevel High-Voltage Pulse Generation Based on a New Modular Solid-State Switch. *IEEE Trans. Plasma Sci.* **2014**, *42*, 2956–2961. [CrossRef]

16. Fang, T.; Ruan, X.; Tse, C.K. Control Strategy to Achieve Input and Output Voltage Sharing for Input-Series-Output-Series-Connected Inverter Systems. *IEEE Trans. Power Electron.* **2010**, *25*, 1585–1596. [CrossRef]

17. Guo, Z.; Sha, D.; Liao, X.; Luo, J. Input-Series-Output-Parallel Phase-Shift Full-Bridge Derived DC–DC Converters with Auxiliary LC Networks to Achieve Wide Zero-Voltage Switching Range. *IEEE Trans. Power Electron.* **2014**, *29*, 5081–5086. [CrossRef]

18. Ahn, H.M.; Sung, W.Y.; Kim, M.K.; Ryu, S.H.; Lim, C.S.; Lee, B.K. Control Method of Input-Parallel and Output-Series Connected Inverters for Plasma Generator. In Proceedings of the 2018 IEEE Applied Power Electronics Conference and Exposition (APEC), San Antonio, TX, USA, 4–8 March 2018; pp. 3563–3568.

19. Alou, P.; Oliver, J.; Cobos, J.A.; Garcia, O.; Uceda, J. Buck+half Bridge (d=50%) Topology Applied to very Low Voltage Power Converters. In Proceedings of the 2001 Sixteenth Annual IEEE Applied Power Electronics Conference and Exposition (2001 APEC), Anaheim, CA, USA, 4–8 March 2001; pp. 715–721.

20. Borage, M.; Tiwari, S.; Kotaiah, S. Analysis and Design of an LCL-T Resonant Converter as a Constant-Current Power Supply. *IEEE Trans. Ind. Electron.* **2005**, *52*, 1547–1554. [CrossRef]

21. Bhat, A.K.S. A Fixed Frequency LCL-Type Series Resonant Converter. *IEEE Trans. Aerosp. Electron. Syst.* **1995**, *31*, 97–103. [CrossRef]

energies

MDPI

Article

Gasification of Waste Cooking Oil to Syngas by Thermal Arc Plasma

Andrius Tamošiūnas [1,*], Dovilė Gimžauskaitė [1], Mindaugas Aikas [1], Rolandas Uscila [1], Marius Praspaliauskas [2] and Justas Eimontas [3]

1   Plasma Processing Laboratory, Lithuanian Energy Institute, Breslaujos str. 3, LT-44403 Kaunas, Lithuania
2   Laboratory of Heat Equipment Research and Testing, Lithuanian Energy Institute, Breslaujos str. 3, LT-44403 Kaunas, Lithuania
3   Laboratory of Combustion Processes, Lithuanian Energy Institute, Breslaujos str. 3, LT-44403 Kaunas, Lithuania
*   Correspondence: Andrius.Tamosiunas@lei.lt; Tel.: +370-37-401-999

Received: 4 June 2019; Accepted: 4 July 2019; Published: 7 July 2019

Abstract: The depletion and usage of fossil fuels causes environmental issues and alternative fuels and technologies are urgently required. Therefore, thermal arc water vapor plasma for a fast and robust waste/biomass treatment is an alternative to the syngas method. Waste cooking oil (WCO) can be used as an alternative potential feedstock for syngas production. The goal of this experimental study was to conduct experiments gasifying waste cooking oil to syngas. The WCO was characterized in order to examine its properties and composition in the conversion process. The WCO gasification system was quantified in terms of the produced gas concentration, the H_2/CO ratio, the lower heating value (LHV), the carbon conversion efficiency (CCE), the energy conversion efficiency (ECE), the specific energy requirements (SER), and the tar content in the syngas. The best gasification process efficiency was obtained at the gasifying agent-to-feedstock (S/WCO) ratio of 2.33. At this ratio, the highest concentration of hydrogen and carbon monoxide, the H_2/CO ratio, the LHV, the CCE, the ECE, the SER, and the tar content were 47.9%, 22.42%, 2.14, 12.7 MJ/Nm^3, 41.3% 85.42%, 196.2 kJ/mol (or 1.8 kWh/kg), and 0.18 g/Nm^3, respectively. As a general conclusion, it can be stated that the thermal arc-plasma method used in this study can be effectively used for waste cooking oil gasification to high quality syngas with a rather low content of tars.

Keywords: waste cooking oil; gasification; thermal arc plasma; water vapor; syngas

---

1. Introduction

Energy and transportation sectors face several issues on a global scale, such as a broad dependence on fossil fuels, their fluctuating prices and environmental impact. A strong global dependence on fossil fuels and their association with greenhouse gas emissions have laid emphasis on the prospects of alternative biofuels at an economically sustainable level. Regarding these goals, there have been significant achievements in the processing of biofuels. Strong attempts of commercialization of biofuels, as well as the development of compatible engines have evolved to advanced levels. Various feedstocks, such as lignocellulosic biomass (forestry residues, agricultural residues and energy crops), wastes (municipal solid waste, sewage sludge, refuse-derived fuels, animal manure, and industrial wastes), and algae, have been tested as sources in pyrolysis, gasification, liquefaction and anaerobic digestion (fermentation) to produce biofuels (biodiesel, bio-oil, bioethanol, biogas, hydrogen and/or syngas) [1–5].

Vegetable oils can also be considered as a potential fuel to substitute fossil fuels. In general, vegetable oil is categorized as edible and non-edible. Both may consist of triglycerides, which are derived from glycerol and the chains of three fatty acids bounded to glycerol by the carbonyl group [6]. Long-time deep-frying at high temperatures leads to the loss of the cooking oil's properties, such as nutritional value, flavour, texture, and formation of carcinogens and other toxic compounds due to hydrolysis, oxidation, isomerization, and polymerization. As a result, the disposal of waste cooking oil (WCO) has become an unattended problem in terms of environmental and human health issues.

The worldwide production of vegetable oils has been continuously growing, reaching about 200 million metric tons (MMT) in 2018. Between 2012 and 2018, the production volume of vegetable oil increased by around 20% from 161.6 MMT to 204 MMT [7]. The consumption of vegetable oils worldwide in 2018 was 197 MMT, which consisted of palm oil (69.57 MMT), soybean oil (57.05), rapeseed oil (27.83 MMT), sunflower seed oil (17.75 MMT), palm kernel oil (7.95 MMT), peanut oil (5.53 MMT), cottonseed oil (5.15 MMT), coconut oil (3.41 MMT) and olive oil (3.07 MMT) [8]. Therefore, enormous amounts of waste cooking oil are generated globally every day at restaurants, pantry services and at massive scale frying and food processing. The annual quantity of WCO production differs in various countries and is around 18.5 million tons. Globally, the leader is the U.S., accounting for 10 million tons/year (or 55% of total production), followed by China (4.5 million tons/year), and the EU-28 (2.6 million tons/year). Japan, Malaysia, Canada and Taiwan also report some production of WCO, but only in much smaller quantities: 0.57, 0.5, 0.12 and 0.07 million tons, respectively [9].

Raw vegetable oil is not suitable for direct application in diesel engines. Problems related to using vegetable oils as engine fuel may result in short-term and long-term problems. The short-term problems are caused by its high viscosity (cold weather starting), the presence of natural gums and ash, as well as a low cetane number causing engine knocking. Long-term usage of vegetable oil may cause the coking of injectors, carbon deposition on pistons and the heads of engines due to the incomplete combustion of fuel. Moreover, failure of engine lubrication may occur due to oil polymerization [10,11].

The transesterification of vegetable oil and animal fats is one of the most commonly used methods, enabling the reduction of its viscosity, in addition to drying and condensation in cold weather. A number of different biodiesel production technologies, from vegetable oils and waste vegetable oils to animal fat, were investigated in [12–15]. However, the major challenge in the production of biodiesel from vegetable oils is the cost of raw materials and their limited availability. Fangrui and Milford [12] report that the cost of oils and fats accounts for 60% to 75% of the cost of biodiesel fuel. Therefore, the use of waste cooking oil can reduce costs, but the quality of used waste cooking oils can be poor. Waste cooking oil can also be a plausible source for the production of biolubricants [16,17].

Despite the production of biodiesel and biolubricants from WCO, it can also serve as a promising precursor for hydrogen and/or synthesis gas (syngas) production via the thermochemical pathway. Young-Doo Kim et al. [18] studied fresh and waste soybean oil gasification with air in a bench-scale fluidised-bed reactor. The effects of the equivalence ratio on the gas composition and tar content, as well as low-temperature oxidation on the fuel properties, were examined. It was stated that low-temperature oxidation of fuel improved the quality of the producer gas as the oleic and stearic acids contents in waste soybean oil increased, while the linoleic acid content decreased. Wu et al. [19] converted waste rapeseed oil to gaseous products (H_2, CO, CH_4, C_2H_2, C_2H_4, C_2H_6) using the aerosol gliding arc discharge plasma in the environment of an argon and oil mixture. It was found that 74% of the waste rapeseed oil was converted into gases at an applied voltage of 10 kV and a temperature of 800 °C. Meier et al. [20] performed WCO thermal cracking (fast pyrolysis) in a continuous tubular reactor under three isothermal temperatures (475, 500 and 525 °C) and different residence time (5–70 s). The dominant gaseous products detected were H_2 and C_3–C_4, light hydrocarbons with concentrations of 45.7% and 48.9%, respectively, at a temperature of 500 °C. Yenumala et al. [21] conducted a thermodynamic equilibrium analysis of steam reforming (SR) and autothermal steam reforming (ASR) vegetable oils to syngas using the Gibbs free energy minimization method. The effects of a broad range of temperatures (573–1273 K) and a steam-to-carbon ratio (1:6) on the H_2 yield and the selectivity of CO and CH_4 were carried out. It was found that the yield of H_2 increased with the increase of the steam-to-carbon ratio and temperature, but the selectivity of CH_4 decreased. The optimum experimental conditions for steam reforming vegetable oil for a maximum H_2 yield and a low selectivity of CH_4 were found to be 875–925 K and a steam-to-carbon ratio of 5–6.

Generally, there is limited scientific literature available on the conversion of waste cooking oil for hydrogen-rich synthesis gas production via the thermochemical route of gasification. Therefore, this study focuses on the gasification of waste cooking oil to syngas using thermal arc plasma. Water vapor was used as the main plasma-forming gas with a small amount of air to protect the cathode from erosion. The gasification of WCO was quantified in terms of the producer gas concentration, the tar content in the producer gas and the condensate, the lower heating value (LHV), the H_2/CO ratio, the carbon conversion efficiency (CCE), the energy conversion efficiency (ECE), and the specific energy requirements (SER).

2. Materials and Methods

2.1. Waste Cooking Oil and Its Characterization Methods

Waste cooking oil was received from a local JSC Bionova in Kaunas, Lithuania. The company collects used cooking oil from restaurants and frying and food processing companies across Lithuania. Generally, the type of used vegetable oil received could not be clearly identified. Therefore, a proximate and ultimate analysis, as well as a gas chromatography method were used to characterise the waste cooking oil. The proximate (volatile matter, fixed carbon, ash, moisture and lower heating value) and the ultimate (content of C, H, N, S, O, Cl) analyses were performed by a fuel elements analyser Flash 2000 (ThermoFisher Scientific, the Netherlands), a fuel calorimeter IKA C5000 (IKA, Germany) and a thermogravimeter TGA 4000 coupled with a gas chromatograph Clarus 680 and a mass spectrometer Clarus 600 T (Perkin Elmer, USA) according to standards LST EN ISO 16948:2015, LST EN ISO 16994:2016 [22]. A gas chromatograph Clarus 500 (Perkin Elmer, USA) was used to measure the composition of the fatty acids in the WCO according to the methodology provided in the standard LST EN ISO5508 [23].

2.2. WCO Gasification Setup

For the gasification of waste cooking oil to syngas, a plasma chemical reactor of 0.0314 m^3 was used. The reactor was insulated with Al_2O_3 ceramics, with a thickness of 25 mm. The length of the reactor was 1 m, with an inner diameter of 0.2 m. The schematic of the experimental setup is shown in Figure 1, which includes the plasma torch (1), the chemical reactor (2), the plasma-forming gas feeding system (3), the electric circuit (4), the waste cooking oil feeding with preheating (5), and a line for producer gas sampling and analysis (6). T1, T2, T3, and T4 stand for the thermocouples used to measure the temperature gradient in the reactor and the producer gas temperature. An R-type thermocouple was used to measure the temperature at the position T1, and a K-type thermocouple was used to measure the temperature at the positions T2, T3, and T4. However, temperature measurement at the position T1 was challenging, as the thermocouple over-exceeded its temperature range, reaching 1430 °C. The rest of the temperature gradient is as follows: 950 °C at T2, 700 °C at T3, and 550 °C at T4. A similar gasification system was used for waste glycerol gasification to syngas [24]. Additional information about the experimental gasification system and the plasma torch can be found in previous research [25–27].

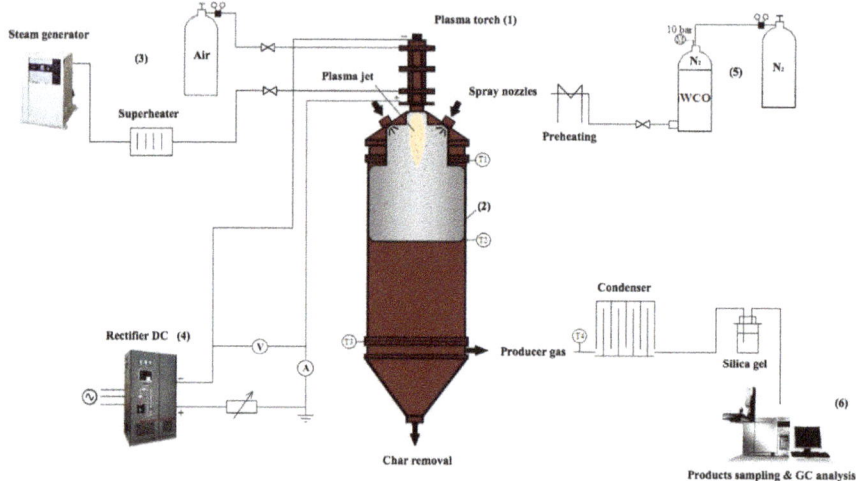

Figure 1. Experimental gasification setup of waste cooking oil.

A linear-type direct current (DC) arc plasma torch with 48–57.6 kW power (I = 160 A, U = 300–360 V) was used to generate plasma stream at atmospheric pressure. The electrodes of the plasma torch were made of copper and a hafnium cathode was used as an electron emitter. The anode of the torch had a stair-step shape in order to minimize the electric arc's pulsations, thus making the operation of the plasmatron more stable. The length of the stair-step anode was 0.18 m with an inner diameter of 12 mm at the narrowest part and 18 mm at the exit nozzle. The length of the narrow part of the anode was 30 mm and the remaining wider part was 150 mm. An electromagnetic coil was also used in order to additionally increase the arc's rotation and thus prolong the lifetime of the anode. The plasma torch was water cooled. Water vapor was used as the main plasma-forming gas, heat carrier and reactant with a small admixture of air (up to 17%) tangentially supplied near the hafnium cathode for erosion protection. A 3 kW superheater was used to overheat the water vapor to 240 °C prior to its supply to the plasma torch. This is an essential condition to avoid condensation on the inner walls of the arc discharge chamber of the plasma torch and thus prolong the lifetime of the electrodes.

The waste cooking oil was preheated to around 110 °C in order to reduce its viscosity, thus ensuring spraying stability, constant flow rate and better atomization. A special commercially available spraying nozzle by the Danfoss company with a 2.21 g/s (7.95 kg/h) capacity was used to inject the waste cooking oil into the plasma chemical reactor. A constant 10 bar pressure was kept in the WCO feeding line in order to guarantee a stable flow rate at the capacity limits of the nozzle. The spraying angle of the nozzle was 60° with an S-type solid spray pattern.

The producer gas was condensed and sent for sampling and analysis. The concentrations of gaseous products were measured by an Agilent 7890 A gas chromatograph (GC) equipped with a dual-channel thermal conductivity detector and a valve system.

2.3. Tar Content Measurement and Formation Mechanism

The measurement of the tar content in the producer gas and the condensate was also performed. The entire procedure was done according to the standard EN 1948-2:2006 [28].

Extraction: tar extraction was performed with toluene at ambient temperature. The formed emulsion was disrupted by the addition of a saturated sodium chloride solution.

Distillation: the flask was filled with liquid and inserted into the rotary distiller. The temperature was about 60 °C and the pressure was 550 mmHg. The speed of the distillation was about 2–3 drops per second.

Analysis: Analysis of the calibrated compounds was performed with a Varian GC-3800 gas chromatograph equipped with a flame ionization detector (FID). Restek RXI-5ms universal 60 m long with a 0.25 mm inner diameter capillary column with a 0.25 μm thick (5% phenol) methylpolysiloxane layer was used for chromatographic separation of the compounds. The main conditions of the experiment are as follows: an injector temperature of −275 °C, a dilution gas ratio of 1:75, and chromatographic column temperatures from 50 °C to 325 °C (8 °C/min). The carrier gas was helium with a 1.2 mL/min gas flow. The compounds obtained during the experiment were identified by the characteristic output times obtained by analyzing the calibration mixture EPA 610.

The formation of tars starts after the linear oil's thermal decomposition to cyclic compounds. At thermal degradation points, linear oils decompose, forming various polycyclic aromatic hydrocarbons (PAHs). The formation of PAHs may have various mechanisms, wherein hydrogen-abstraction acetylene addition is widely acknowledged. Below, a simple illustration of naphthalene generation from benzene through intermediate phenyl acetylene is shown [29].

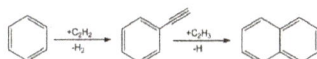

2.4. Quantification Parameters

In order to assess the performance of any gasification system, basic measure parameters were used. These measure parameters include the concentrations of compounds present in producer gas, the syngas yield, the H_2/CO ratio, the LHV, the carbon conversion efficiency (CCE), the energy conversion efficiency (ECE), and the specific energy requirements (SER). The methodology enabling the evaluation of the performance of the gasification system is discussed in more detail in [24,26].

The LHV (MJ/Nm^3):

$$LHV = 10.78H_2(\%) + 12.63CO(\%) + 35.88CH_4(\%) + 56.5C_2H_2(\%) + 64.5C_2H_6(\%) + 93.21C_3H_8(\%), \quad (1)$$

where H_2 (%), CO (%), CH_4 (%), C_2H_2 (%), C_2H_6 (%), and C_3H_8 (%) are the content of gaseous products in the producer gas (vol.%).

The CCE (%):

$$CCE = 12 \times Y_{dry\ gas} \times \left\{ \frac{[CO + CO_2 + CH_4] + 2 \times [C_2H_2 + C_2H_6] + 3 \times C_3H_8}{22.4 \times C} \right\} \times 100\%, \quad (2)$$

where $Y_{dry\ gas}$ is the dry gas yield of dry feedstock (Nm^3/kg), CO, CO_2, CH_4, C_2H_2, and C_2H_6 are in v/v (%), and C is in wt.% of carbon in dry feedstock.

The ECE (%):

$$ECE = \frac{m_{syngas} \times LHV_{syngas}}{P_{plasma} + m_{WCO} \times LHV_{WCO}} \times 100\%, \quad (3)$$

where m_{syngas} and m_{WCO} are the mass flow rates of syngas and feedstock (kg/s), respectively. LHV_{syngas} and LHV_{WCO} are the net calorific values of the produced syngas and the feedstock (MJ/kg), respectively. P is the plasma torch power (kW).

The SER (kJ/mol or kWh/kg):

$$SER = \frac{P_{plasma}}{m_{syngas}}, \quad (4)$$

where SER is the specific energy requirement (kJ/mol or kWh/kg), P_{plasma} is the plasma torch power (kJ/s), and m_{syngas} is the mass flow rate of produced syngas (mol/s).

3. Results

3.1. WCO Characterisation

Tables 1 and 2 present the ultimate and proximate analysis of the waste cooking oil used as a feedstock for gasification as well as the fatty acid composition in it, respectively. As can be seen from Table 1, the WCO was mostly composed of carbon, hydrogen and oxygen compounds with traces of nitrogen, sulphur and chlorine. The lower heating value of the WCO was determined to be 39.24 MJ/kg and the volatile organic matter comprised 99.15 wt.%.

Table 1. Ultimate and proximate analysis of the waste cooking oil (WCO).

Parameter	WCO	Standard
Ultimate analysis, wt.%		
C	71.84 ± 2.99	
H	10.14 ± 2.11	
N	0.06 ± 0.003	
S	<0.01 (0.008)	
O *	17.71	
Cl	0.003	LST EN ISO 16948:2015
Proximate analysis, wt.%		LST EN ISO 16994:2016
VOCs	99.15 ± 1.0	
Fixed carbon	0.56 ± 0.003	
Ash	0.24 ± 0.004	
Water content	0.08	
Lower heating value, MJ/kg	39.24 ± 0.03	

* by difference.

Table 2. Fatty acid composition of the WCO.

Fatty Acids	Structure [a]	Formula	Composition (wt.%)	Detection Method
Myristoleic	C14:1	$C_{14}H_{26}O_2$	0.26 ± 0.008	
Pentadecanoic	C15:0	$C_{15}H_{30}O_2$	0.04 ± 0.004	
Palmitic	C16:0	$C_{16}H_{32}O_2$	6.85 ± 0.041	
Palmitoleic	C16:1	$C_{16}H_{30}O_2$	0.23 ± 0.016	
Stearic	C18:0	$C_{18}H_{36}O_2$	2.36 ± 0.037	
Oleic	C18:1	$C_{18}H_{34}O_2$	54.44 ± 0.775	LST EN ISO5508
Linoleic	C18:2	$C_{18}H_{32}O_2$	27.08 ± 0.114	
Linolenic	C18:3	$C_{18}H_{30}O_2$	5.96 ± 0.049	
Arahidic	C20:0	$C_{20}H_{40}O_2$	0.86 ± 0.0082	
Eikosenic	C20:1	$C_{20}H_{38}O_2$	1.00 ± 0.008	
Lignoceric	C24:0	$C_{24}H_{48}O_2$	0.27 ± 0.041	
Insoluble impurities in the WCO [b]	-	-	6.31 ± 0.12	LST EN ISO 663

[a] xx:y indicates xx carbons in the fatty acid chain with y number of double bonds. [b] Insoluble impurities detected in the WCO, including oxidized fatty acids.

Since vegetable oils are considered as a mixture of triglycerides, i.e., an ester consisting of glycerol and three fatty acids, saturated or unsaturated aliphatic hydrocarbon compounds present in the vegetable oils generally vary from 8 to 24 carbon atoms. The dominant majority of carbon atoms usually varies between 16 and 18 [21]. As can be seen from Table 2, the dominant carbon atoms in the WCO used for gasification to syngas ranged from 16 to 18. Palmitic, stearic, oleic, linoleic, and linolenic acids were the major fatty acids constituting the waste cooking oil used as a feedstock, which comprised more than 90 wt.%. It was also detected that insoluble impurities, including oxidized fatty acids, were present in the WCO, amounting to 6.31 wt.%.

3.2. Effect of Gasifying the Agent-to-Feedstock Ratio on the Gasification Efficiency of Waste Cooking Oil

Water vapor was used as the gasifying agent, but it also served as the main plasma-forming gas and heat carrier, constituting a generated plasma stream with active radicals inside. The flow rate of the water vapor entering the plasma torch to be heated by the electric arc to form the plasma stream ranged from 2.4 g/s to 4.65 g/s, whereas the flow rate of the WCO was kept constant at 2.21 g/s. Generally, the change in the flow rate of water vapor is directly linked with the change in the power of the plasma torch through the relation $P = IU$. At the constant arc's current intensity, the change in the flow rate of plasma-forming gas may increase or decrease the arc voltage. As a result, the power of the plasma torch changes. The change in the power of the plasma torch is also directly coupled with the change in the enthalpy of the generated plasma stream, and thus the temperature, due to the relation $h_f = f(T_f)$. Therefore, the effect of the water vapor-to-waste cooking ratio on the gasification process efficiency was determined. For simplicity, this ratio was chosen as a principle and is noted as an S/WCO ratio. It should be kept in mind that this ratio also corresponds (is equal) to the power-to-WCO ratio or the temperature-to-WCO ratio, as the trend of the curves in the figures remains the same. The mean temperature of the generated plasma stream entering the plasma chemical reactor was in the range of 2600–3000 ± 50 K.

The effect of the S/WCO ratio on the producer gas concentrations is shown in Figure 2. As can be seen from the figure, the prevailing producer gases were hydrogen and carbon monoxide, followed by carbon dioxide and methane, with highest concentrations of 47.9%, 22.42%, 7.74%, 7.83%, respectively, at the S/WCO ratio of 2.33. Moreover, higher hydrocarbons were also observed, such as acetylene (C_2H_2), ethane (C_2H_6) and propane (C_3H_8), with concentrations of 2.27%, 0.42% and 0.37%, respectively, at the same ratio. As the S/WCO ratio increased from 1.31 to 2.33 (the plasma torch power increased from 48 kW to 57.6 kW), the concentrations of the constituents in the producer gas did not change significantly. The highest increase was observed only for H_2 when the concentration increased from 40.58% to 47.9%. This is mostly attributed to the dominance of steam reforming, water–gas shift (WGS) and cracking reactions [24]. The concentration of CO was almost stable at any S/WCO ratio and was around 22%–23.5%. During the WCO conversion to syngas process, the observed concentration of CH_4 was rather high, up to 9.44%. The increased S/WCO ratio, from 1.31 to 2.33, led to a slight decrease in the methane content, from 9.44% to 7.83%. This can be explained by reverse methanation and hydrogenation reactions, whereby methane and water forms hydrogen, carbon monoxide or carbon dioxide. The concentration of the latter at this ratio increased from 5.83% to 7.74%. Additionally, higher hydrocarbons, such as acetylene, ethane and propane, formed due to insufficient residence time to fully convert the carboxylic acids with long aliphatic chains to simple H_2, CO and CO_2 molecules.

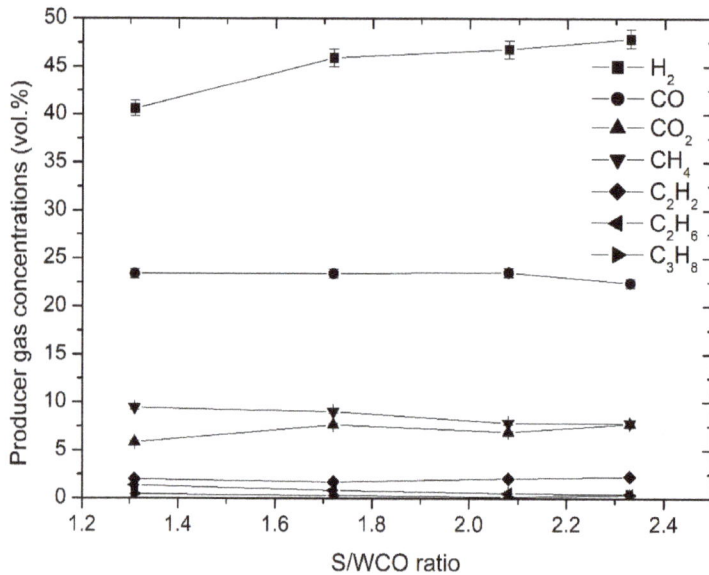

Figure 2. Effect of the S/WCO ratio on the producer gas composition.

The effect of the S/WCO ratio on the quality of the produced syngas and the H_2/CO ratio is shown in Figure 3. The lower heating value of the producer gas (syngas) did not change significantly over the variable range of the S/WCO ratio during the experiments. The LHV was in the range of 12.5 MJ/Nm^3 to 13.2 MJ/Nm^3, which shows the production of good quality syngas. This was mostly attributed to the steam reforming reaction. Since water vapor was simultaneously used as a plasma-forming gas, a heat carrier and a gasifying agent, an additional portion of hydrogen came from the water-splitting reaction to hydrogen and oxygen, induced by its passage through the high temperature electric arc discharge region. According to the thermodynamic equilibrium calculations, the content of hydrogen coming from the water vapor plasma may constitute up to 9–10% (vol.) at the mean plasma stream temperature of 2800–3000 K [30]. The obtained H_2/CO ratio of 2 showed the potential of the produced syngas to be directly used for biofuels production via Fischer–Tropsch (FT) synthesis without any ratio adjustment by the WGS reaction. The only problem might be the requirement of tar removal from the syngas prior to the application of the reaction. In the case of the syngas to be used for methane production via hydrogen methanation with carbon monoxide (5) or carbon dioxide (6), the process would require a H_2/CO or H_2/CO_2 ratio adjustment. To perform a methanation reaction (5), the stoichiometric $CO:H_2$ ratio should be 1:3. Therefore, in this experimental case, an extra hydrogen molecule would be required to ensure the success of the reaction. A methanation reaction (6) could be of greater interest, because the H_2/CO_2 ratio was around 6 to 7, when the theoretically required $CO_2:H_2$ stoichiometric ratio should be at least 1:4. In this case, like in the FT synthesis, tar removal prior to syngas upgrading to methane should be ensured. The tar content in the syngas will be discussed in the next section.

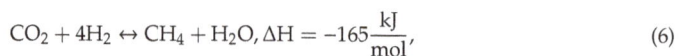

$$CO + 3H_2 \leftrightarrow CH_4 + H_2O, \Delta H = -206\frac{kJ}{mol}, \tag{5}$$

$$CO_2 + 4H_2 \leftrightarrow CH_4 + H_2O, \Delta H = -165\frac{kJ}{mol}, \tag{6}$$

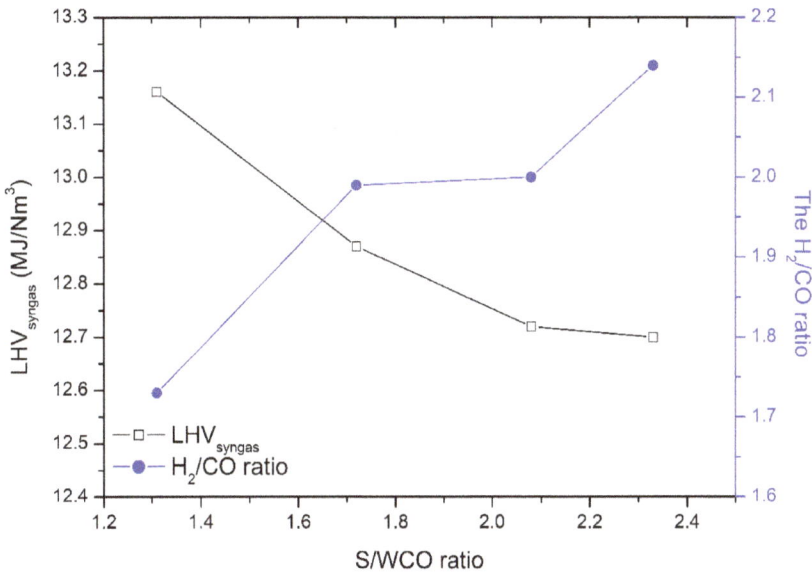

Figure 3. Effect of the S/WCO ratio on the producer syngas quality and the H_2/CO ratio.

The effect of the S/WCO ratio on the carbon conversion efficiency is shown in Figure 4. The highest value of the CCE was obtained at the S/WCO ratio of 2.33 and exceeded 41.3%. As a result, the carbon present in the waste cooking oil was not fully converted into gaseous products. Some solid carbon deposition was observed during the experiments. Due to the presence of 18 carbon atoms in the WCO, the cracking of the long-chain hydrocarbons into smaller and simpler ones takes time. Therefore, some amounts of higher gaseous hydrocarbons as well as solid carbon were obtained.

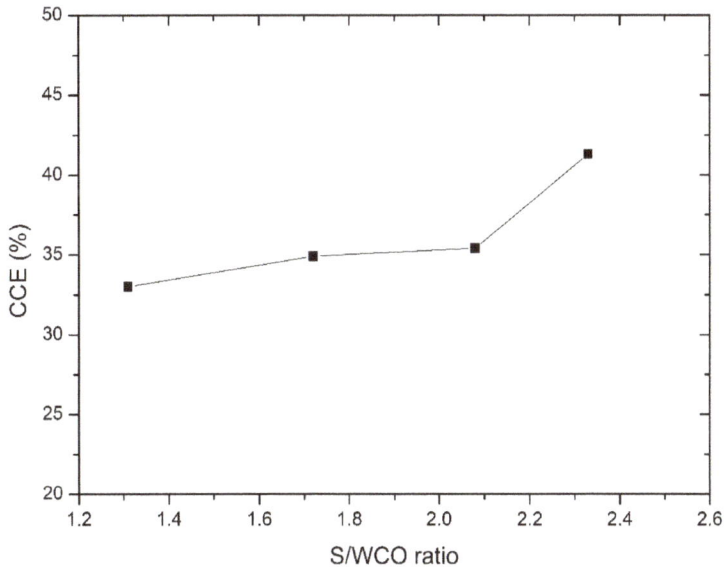

Figure 4. Effect of the S/WCO ratio on the carbon conversion efficiency (CCE).

The effect of the S/WCO ratio on the gasification process performance in terms of the energy conversion efficiency (ECE) and the specific energy requirements (SER) is shown in Figure 5. As can be seen from the figure, the increased S/WCO ratio from 1.31 to 2.33 led to an increase in the ECE and a decrease in the SER from 51.54% to 85.42% and 294.7 kJ/mol (2.73 kWh/kg) to 196.2 kJ/mol (1.82 kWh/kg), respectively. As the power of the plasma torch increased from 48 kW to 57.6 kW due to the increased flow rate of the water vapor, this should lead to a decrease in the energy conversion efficiency. However, more syngas was produced at the constant flow rate of feedstock (Equation (3)) which compensated for the energy losses and led to an increase in the ECE. All the spent energy for the process should be included in the optimized process case, which would lower the value of the ECE. This should include the energy consumed during steam production, steam superheating, feedstock heating, etc., which is not currently optimized. In addition, the utilization of the process waste heat would help to reduce the overall losses and increase the ECE. The specific energy needed to convert one mole of the waste cooking oil to syngas decreased, thus making the process more efficient, even if the power consumed to run the plasma torch increased. This was mostly attributed to the higher syngas production volume (Equation (4)).

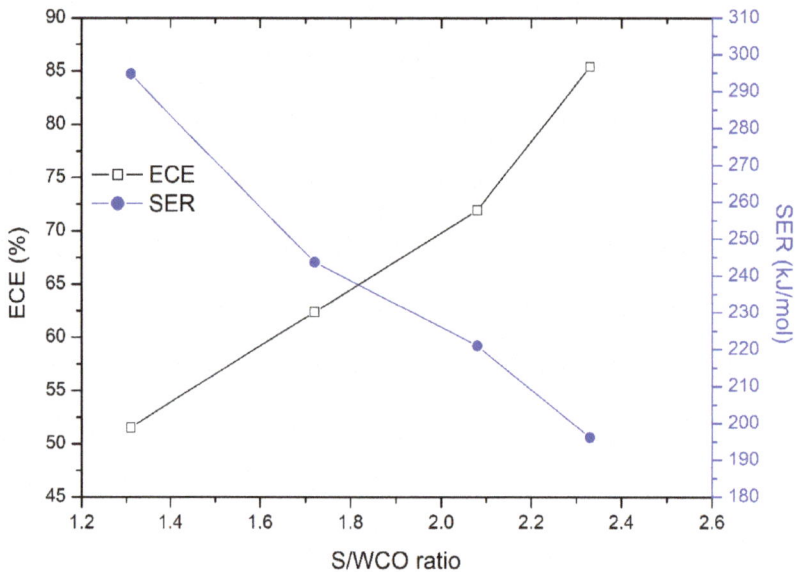

Figure 5. Effect of the S/WCO ratio on the energy conversion efficiency (ECE) and the specific energy requirements (SER).

3.3. Tar Content in the Producer Gas (Syngas)

The tar content in the producer gas, as well as in syngas, is an important parameter showing how clean the produced syngas is and what kind of methods should be chosen in case of a syngas upgrade to biodiesel, methane, hydrogen or direct burning. The effect of the S/WCO ratio on the tar concentration in the producer gas (syngas) is shown in Figure 6 and the main compounds present in the tar are shown in Table 3.

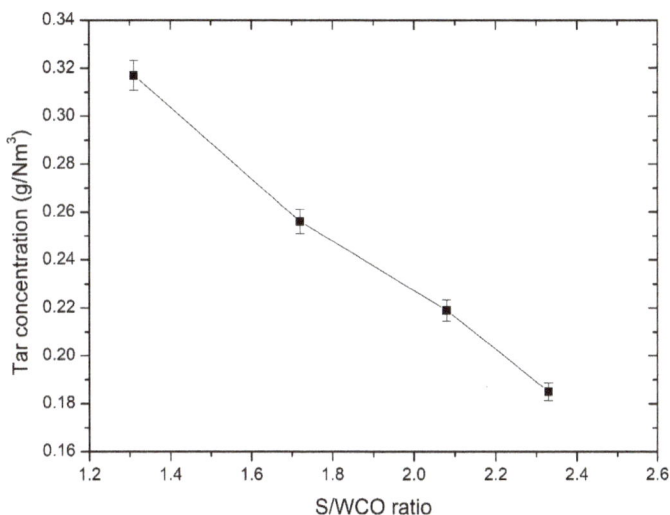

Figure 6. Effect of the S/WCO ratio on the tar content in the producer gas (syngas).

Table 3. Composition of tars detected in the syngas at the ratio of S/WCO 1.31 and 2.33.

Identified Compound	Concentration in Syngas (g/Nm³)	
	S/WCO—1.31	S/WCO—2.33
Naphthalene	0.222	0.13
Fluorene	0.022	0.012
Fluoranthene	0.022	0.013
Pyrene	0.020	0.012
Benzo[ghi]perylene	0.019	0.0011
Antracene	0.008	0.005
Phenanthrene	0.003	0.002
Total:	0.317	0.185

A qualitative and quantitative analysis of the main tar compounds obtained in the syngas shows that naphthalene was present as the main polycyclic aromatic compound with the highest concentration exceeding around 70% of the total concentration. The remaining identified compounds, such as fluorine, fluoranthene, pyrene, benzo[ghi]perylene, antracene and phenanthrene all comprised up to 30% of the total concentration defined in the tar. The highest concentration of tar in the producer gas was obtained at the S/WCO ratio of 1.31 and was 0.317 g/Nm³, whereas the lowest content of 0.185 g/Nm³ was found at the S/WCO ratio of 2.33. Normally, if syngas is used in an internal combustion engine (ICE) for power generation, the tar content should be lower than 0.1 g/Nm³ [31]. Therefore, the produced syngas quality in this experimental research would require additional tar removal or process optimization if the syngas was planned to be used for power generation in the ICE. Kim et al. [18] used the activated carbon filter installed just after the electrostatic precipitator in the treatment of the WCO to syngas in a fluidised-bed reactor. It was reported that the filter did not affect the concentration of the syngas, but the tar content decreased significantly from 0.308 g/Nm³ to 0.069 g/Nm³. Besides the tar content, the concentration of solid particles in the producer gas is also an important parameter. The limit value of the particulates must be lower than 50 mg/Nm³ for the ICE to be powered by syngas for electricity production. However, this parameter was not determined in this study. Various gas cleaning systems might be used for tar and particle removal, as reported in [32]. Generally, the use of the plasma method may also help to reduce the tar and particulate content in the product gases without a need for filtering, catalyzing or other cleaning methods [33].

4. Conclusions

In this experimental study, waste cooking oil was gasified to syngas using the DC thermal arc-plasma method. Water vapor was simultaneously used as a gasifying agent, a heat carrier and a reactant. The gasification system was quantified in terms of several basic parameters, such as the concentration of the producer gas, the H_2/CO ratio, the LHV, the CCE, the ECE, and the SER. Additionally, the tar content in the product gas was also determined. The findings show that the best process efficiency was achieved at the S/WCO ratio of 2.33, 57.6 kW plasma torch power and the mean plasma stream temperature of 2800 K. In these conditions, the content of syngas in the producer gas was around 70% (H_2—47.9% and CO—22.42%) with some amounts of carbon dioxide (7.74%), methane (7.83%), acetylene (2.27%), ethane (0.42%), and propane (0.37%). The H_2/CO ratio was 2.14, indicating that the produced syngas can be used for biofuels production via FT synthesis. The lower heating value of 12.7 MJ/Nm^3 shows that the syngas was of a high calorific value. The highest CCE was around 41.3%, indicating that the WCO was not fully converted to gaseous products. The ECE and the SER were calculated to be 85.42% and 196.2 kJ/mol (or 1.18 kWh/kg), respectively. The tar content in the producer gas exceeded 0.18 g/Nm^3.

As a general conclusion, it can be stated that the thermal arc-plasma method used in this study can be effectively applied for waste cooking oil conversion to high quality syngas with a rather moderate content of tars.

Author Contributions: Conceptualization, A.T.; methodology, D.G., M.P., J.E.; resources, M.A., R.U.; formal analysis, A.T., R.U., M.A.; writing—original draft preparation, A.T.; supervision, A.T.

Funding: This research received no external funding.

Conflicts of Interest: The authors declare no conflict of interest.

References

1. Ro, D.; Shafaghat, H.; Jang, S.H.; Lee, H.W.; Jung, S.C.; Jae, J.; Cha, J.S.; Park, Y.K. Production of an upgraded lignin-derived bio-oil using the clay catalysts of bentonite and olivine and the spent FCC in a bench-scale fixed bed pyrolyzer. *Environ. Res.* **2019**, *172*, 658–664. [CrossRef] [PubMed]
2. Kim, J.Y.; Lee, H.W.; Lee, S.M.; Jae, J.; Park, Y.K. Overview of the recent advances in lignocellulose liquefaction for producing biofuels, bio-based materials and chemicals. *Bioresour. Technol.* **2019**, *279*, 373–384. [CrossRef] [PubMed]
3. Nanda, S.; Azargohar, R.; Dalai, A.K.; Kozinski, J.A. An assessment on the sustainability of lignocellulosic biomass for biorefining. *Renew. Sustain. Energy Rev.* **2015**, *50*, 925–941. [CrossRef]
4. Tamošiūnas, A.; Valatkevičius, P.; Gimžauskaitė, D.; Valinčius, V.; Jeguirim, M. Glycerol steam reforming for hydrogen and synthesis gas production. *Int. J. Hydrog. Energy* **2017**, *42*, 12896–12904. [CrossRef]
5. Mathimani, T.; Pugazhendhi, A. Utilization of algae for biofuel, bio-products and bio-remediation. *Biocatal. Agric. Biotechnol.* **2019**, *17*, 326–330. [CrossRef]
6. Nanda, S.; Rana, R.; Hunter, H.N.; Fang, Z.; Dalai, A.K.; Kozinski, J.A. Hydrothermal catalytic processing of waste cooking oil for hydrogen-rich syngas production. *Chem. Eng. Sci.* **2019**, *195*, 935–945. [CrossRef]
7. Statista. Available online: https://www.statista.com/statistics/263933/production-of-vegetable-oils-worldwide-since-2000/ (accessed on 15 May 2019).
8. Statista. Available online: https://www.statista.com/statistics/263937/vegetable-oils-global-consumption/ (accessed on 15 May 2019).
9. Yaakob, Z.; Mohammad, M.; Alherbawi, M.; Alam, Z.; Sopian, K. Overview of the production of biodiesel from Waste cooking oil. *Renew. Sustain. Energy Rev.* **2013**, *18*, 184–193. [CrossRef]
10. Sonthalia, A.; Kumar, N. Hydroprocessed vegetable oil as a fuel for transportation sector: A review. *J. Energy Inst.* **2019**, *92*, 1–17. [CrossRef]
11. Meher, L.C.; Vidya Sagar, D.; Naik, S.N. Technical aspects of biodiesel production by transesterification—A review. *Renew. Sustain. Energy Rev.* **2006**, *10*, 248–268. [CrossRef]
12. Fangrui, M.; Milford, H. Biodiesel production: A review. *Bioresour. Technol.* **1999**, *70*, 1–15.

13. Poudel, J.; Karki, S.; Sanjel, N.; Shah, M.; Oh, S.C. Comparison of biodiesel obtained from virgin cooking oil and waste cooking oil using supercritical and catalytic transesterification. *Energies* **2017**, *10*, 546. [CrossRef]

14. Costarrosa, L.; Leiva-Candia, D.E.; Cubero-Atienza, A.J.; Ruiz, J.J.; Dorado, M.P. Optimization of the transesterification of waste cooking oil with mg-al hydrotalcite using response surface methodology. *Energies* **2018**, *11*, 302. [CrossRef]

15. Leung, D.Y.C.; Guo, Y. Transesterification of neat and used frying oil: Optimization for biodiesel production. *Fuel Process. Technol.* **2006**, *87*, 883–890. [CrossRef]

16. Wang, E.; Ma, X.; Tang, S.; Yan, R.; Wang, Y.; Riley, W.W.; Reaney, M.J.T. Synthesis and oxidative stability of trimethylolpropane fatty acid triester as a biolubricant base oil from waste cooking oil. *Biomass Bioenergy* **2014**, *66*, 371–378. [CrossRef]

17. Chowdhury, A.; Chakraborty, R.; Mitra, D.; Biswas, D. Optimization of the production parameters of octyl ester biolubricant using Taguchi's design method and physico-chemical characterization of the product. *Ind. Crops Prod.* **2014**, *52*, 783–789. [CrossRef]

18. Kim, Y.D.; Jung, S.H.; Jeong, J.Y.; Yang, W.; Lee, U. Do Production of producer gas from waste cooking oil in a fluidized bed reactor: Influence of low-temperature oxidation of fuel. *Fuel* **2015**, *146*, 125–131. [CrossRef]

19. Wu, A.; Li, X.; Yan, J.; Zhu, F.; Lu, S. Conversion of the waste rapeseed oil by aerosol gliding arc discharge-assisted pyrolysis. *Int. J. Hydrog. Energy* **2016**, *41*, 2222–2229. [CrossRef]

20. Meier, H.F.; Wiggers, V.R.; Zonta, G.R.; Scharf, D.R.; Simionatto, E.L.; Ender, L. A kinetic model for thermal cracking of waste cooking oil based on chemical lumps. *Fuel* **2015**, *144*, 50–59. [CrossRef]

21. Yenumala, S.R.; Maity, S.K. Reforming of vegetable oil for production of hydrogen: A thermodynamic analysis. *Int. J. Hydrog. Energy* **2011**, *36*, 11666–11675. [CrossRef]

22. Praspaliauskas, M.; Pedišius, N.; Striuigas, N. Elemental Migration and Transformation from Sewage Sludge to Residual Products during the Pyrolysis Process. *Energy Fuels* **2018**, *32*, 5199–5208. [CrossRef]

23. Makarevičiene, V.; Lebedevas, S.; Rapalis, P.; Gumbyte, M.; Skorupskaite, V.; Žaglinskis, J. Performance and emission characteristics of diesel fuel containing microalgae oil methyl esters. *Fuel* **2014**, *120*, 233–239. [CrossRef]

24. Tamošiūnas, A.; Gimžauskaitė, D.; Uscila, R.; Aikas, M. Thermal arc plasma gasification of waste glycerol to syngas. *Appl. Energy* **2019**, *251*, 113306. [CrossRef]

25. Tamošiūnas, A.; Valatkevičius, P.; Grigaitienė, V.; Valinčius, V. Operational parameters of thermal water vapor plasma torch and diagnostics of generated plasma jet. *Rom. Rep. Phys.* **2014**, *66*, 1125–1136.

26. Tamošiunas, A.; Valatkevičius, P.; Grigaitiene, V.; Valinčius, V.; Striugas, N. A cleaner production of synthesis gas from glycerol using thermal water steam plasma. *J. Clean. Prod.* **2016**, *130*, 187–194. [CrossRef]

27. Tamošiūnas, A.; Valatkevičius, P.; Gimžauskaitė, D.; Jeguirim, M.; Mėčius, V.; Aikas, M. Energy recovery from waste glycerol by utilizing thermal water vapor plasma. *Environ. Sci. Pollut. Res.* **2017**, *24*, 10030–10040. [CrossRef] [PubMed]

28. Striugas, N.; Zakarauskas, K.; Stravinskas, G.; Grigaitiene, V. Comparison of steam reforming and partial oxidation of biomass pyrolysis tars over activated carbon derived from waste tire. *Catal. Today* **2012**, *196*, 67–74. [CrossRef]

29. Shukla, B.; Koshi, M. Comparative study on the growth mechanisms of PAHs. *Combust. Flame* **2011**, *158*, 369–375. [CrossRef]

30. Tamošiūnas, A.; Valatkevičius, P.; Valinčius, V.; Grigaitienė, V. Production of synthesis gas from propane using thermal water vapor plasma. *Int. J. Hydrog. Energy* **2014**, *39*, 2078–2086.

31. Hasler, P.; Nussbaumer, T. Gas cleaning for IC engine applications from fixed bed biomass gasification. *Biomass Bioenergy* **1999**, *16*, 385–395. [CrossRef]

32. Han, J.; Kim, H. The reduction and control technology of tar during biomass gasification/pyrolysis: An overview. *Renew. Sustain. Energy Rev.* **2008**, *12*, 397–416. [CrossRef]

33. Striūgas, N.; Valinčius, V.; Pedišius, N.; Poškas, R.; Zakarauskas, K. Investigation of sewage sludge treatment using air plasma assisted gasification. *Waste Manag.* **2017**, *64*, 149–160. [CrossRef] [PubMed]

energies

MDPI

Article

Plasma–Chemical Hybrid NOx Removal in Flue Gas from Semiconductor Manufacturing Industries Using a Blade-Dielectric Barrier-Type Plasma Reactor

Haruhiko Yamasaki *, Yuki Koizumi, Tomoyuki Kuroki and Masaaki Okubo

Department of Mechanical Engineering, Osaka Prefecture University, 1-1 Gakuen-cho, Naka-ku, Sakai, Osaka 599-8531, Japan
* Correspondence: hyamasaki@me.osakafu-u.ac.jp; Tel.: +81-72-254-9233

Received: 28 June 2019; Accepted: 12 July 2019; Published: 16 July 2019

Abstract: NO_x is emitted in the flue gas from semiconductor manufacturing plants as a byproduct of combustion for abatement of perfluorinated compounds. In order to treat NO_x emission, a combined process consisting of a dry plasma process using nonthermal plasma and a wet chemical process using a wet scrubber is performed. For the dry plasma process, a dielectric barrier discharge plasma is applied using a blade-barrier electrode. Two oxidation methods, direct and indirect, are compared in terms of NO oxidation efficiency. For the wet chemical process, sodium sulfide (Na_2S) is used as a reducing agent for the NO_2. Experiments are conducted by varying the gas flow rate and input power to the plasma reactor, using NO diluted in air to a level of 300 ppm to simulate exhaust gas from semiconductor manufacturing. At flow rates of ≤5 L/min, the indirect oxidation method verified greater removal efficiency than the direct oxidation method, achieving a maximum NO conversion rate of 98% and a NO_x removal rate of 83% at 29.4 kV and a flow rate of 3 L/min. These results demonstrate that the proposed combined process consisting of a dry plasma process and wet chemical process is promising for treating NO_x emissions from the semiconductor manufacturing industry.

Keywords: nonthermal plasma; NO_x reduction; PFC; sodium sulfide; wet scrubber; blade-barrier electrode; semiconductor manufacturing

---

1. Introduction

Since 1950, the semiconductor manufacturing industry has continued to grow, raising concerns about the increasing emissions of perfluorinated compounds (PFCs) [1]. PFCs are characterized by a very high global-warming potential, meaning that they produce a large environmental impact, even at low concentrations. As a result, the Kyoto Protocol (COP3) in 1997 set country-specific targets for reducing PFCs. In 2018, COP24 established guidelines for certain regulations on PFCs in developing countries, highlighting the urgency of the task of eliminating PFCs. Methods for eliminating PFC gas used in the semiconductor manufacturing industry include combustion methods, which burn the PFC gas together with such fuels as oil and natural gas to decompose it into hydrogen fluoride and carbon dioxide [2,3], and catalytic methods, which use catalysts to decompose the gas into hydrogen fluoride and carbon dioxide by means of the hydrolysis reaction [4,5]. The hydrogen fluoride generated during treatment is dealt with by dissolving it in water and processing it as hydrogen fluoride solution, or by causing it to react with calcium hydroxide to immobilize it as CaF_2 [6]. Because the catalytic and CaF_2 fixing methods are not suitable for large-scale treatment of PFCs, large-scale treatment facilities have used the combustion method followed by wastewater processing. In the combustion method, NO_x ($NO + NO_2$) is generated as a byproduct of the decomposition of the PFCs. NO_x, which is generated mainly by the combustion of fuel, is a harmful substance that causes photochemical oxidation processes and other chemical changes in the atmosphere and produces aerosols, such as nitric acid (HNO_3), that cause acid rain. Selective catalytic reduction (SCR) is a common technique for processing NO_x. In the combustion method, however, immediately after the PFCs are burned, a water spray is applied to dissolve the resulting hydrogen fluoride (HF) in water. The problem for SCR is that the water spray brings the target exhaust gas down to a normal temperature, at which the catalyst cannot be activated. Furthermore, the HF can sometimes poison the catalyst for the SCR [7,8], which poses a problem for the treatment of the NO_x generated during processing of the PFCs.

Plasma–chemical methods that use nonthermal plasma have been proposed as a method of removing the NO_x at low temperatures [9–14]. The plasma–chemical method oxidizes NO to NO_2 by using nonthermal plasma to promote oxidation, and then it applies a scrubber to reduce the NO_2 to N_2. Our research group has achieved highly efficient removal of NO_x through treatment by a combined process of indirect oxidation by nonthermal plasma together with a sodium sulfite (Na_2SO_3) scrubber [9–12]. Kim et al. have also achieved NO_x removal efficiency of more than 80% using nonthermal plasma–chemical methods to treat the NO_x discharged from semiconductor plants [15]. Comparing the NO_x removal efficiency of two different reducing agents, Na_2S and Na_2SO_3, Kim et al. found that Na_2S was highly efficient at removing NO_x at 1/10 the concentration of Na_2SO_3 (0.1 mass%). Although the cost of using Na_2S is approximately three times higher than that of using Na_2SO_3, using Na_2S as the reducing agent suppresses the reaction of the exhaust gas with oxygen, making high-efficiency treatment of NO_x possible at 1/10 concentration (0.1 mass%) and reducing the cost of treating NO_x to one-third the cost of Na_2SO_3 [15].

With respect to dielectric barrier discharge (DBD) plasma excitation, research has been conducted on the effect of changing the shape of the electrode on its DBD discharge characteristics and ozone-generating characteristics [16]. This research shows that the discharge energy efficiency and ozone generation efficiency are improved by using a multipoint electrode or trench electrode rather than a plate electrode. In concentric cylindrical plasma reactors, which are commonly used in environmental plasma processing, the high voltage is applied perpendicular to the direction of the gas flow, which means the voltage value imposes limits on the possible cross-sectional area or distance between electrodes. Parallelization is therefore required to increase the processing flow rate. However, in the blade-barrier electrode-based plasma reactor proposed in this study, the discharge direction is parallel to the gas flow, making it suitable for high-flow treatment of exhaust gas with no limit on the flow rate, and promising a highly efficient treatment of exhaust gas.

As mentioned above, previous studies have evaluated individual performance such as NO_x reduction by chemical scrubber and NO oxidation by ozone generator, but the performance of the plasma-chemical hybrid technique has not been clarified. In the present study, a method for treating NO_x is considered that combines the nonthermal plasma oxidation method at normal temperatures with a wet scrubber. For the plasma process, there are two different plasma oxidation methods: a direct oxidation method, which irradiates plasma directly into the exhaust gas, and an indirect oxidation method, which combines the exhaust with radical gas irradiated with plasma in the air. NO in the simulated exhaust gas is oxidized to NO_2 by using a blade-barrier electrode reactor as a plasma reactor, and then NO_x reduction is performed by Na_2S scrubber.

2. Principle

In this section, the principle underlying the removal of NO_x by means of a combined process consisting of a dry plasma process using nonthermal plasma and wet chemical process using a wet scrubber is described. The nonthermal plasma causes some of the oxygen present in the air to change into ozone (which is an oxidizer) and oxygen radical (O). Although some of the NO is reduced to N_2 by the reactions indicated in Equations (1) and (2) [17], most of the NO is oxidized to NO_2 by the ozone and radical oxygen, as shown in Equations (3) and (4) [17],

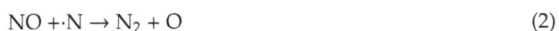

$$N_2 + e- \rightarrow \cdot N + \cdot N + e \tag{1}$$

$$NO + \cdot N \rightarrow N_2 + O \tag{2}$$

$$NO + O_3 \rightarrow NO_2 + O_2 \tag{3}$$

$$NO + O + M \rightarrow NO_2 + M \tag{4}$$

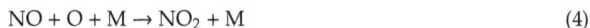

where N and O are radicals, e- is an electron, M is the third-body substance, and N_2 and O_2 are molecules in the air. Reaction (4) proceeds in the presence of oxygen radicals in the plasma. Because the NO_2 obtained by oxidation reactions (3) and (4) is water-soluble, by using the Na_2S scrubber after oxidation, the NO_2 is reduced to N_2 and Na_2SO_4, which are nontoxic and water-soluble. This reaction is shown in Equation (5).

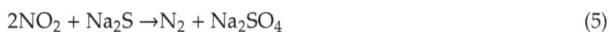

$$2NO_2 + Na_2S \rightarrow N_2 + Na_2SO_4 \tag{5}$$

3. Experimental Apparatus and Method

In this study, a blade-barrier electrode plasma reactor is used to generate ozone and radical oxygen for the purpose of oxidizing the NO, as shown in Equations (3) and (4). A schematic diagram of the blade-barrier plasma reactor and a detailed view of the blade electrode are shown in Figure 1. To increase the strength of the electric field, a multihole ($\varphi 2 \times 16$) blade electrode is used. The gas flows from the top of the reactor, passes through the multihole blade electrode, and enters the discharge part. After the reaction in the discharge part, the gas flows out of the reactor through the holes in the lower electrode ($\varphi 2 \times 4$). To elicit DBD discharge, a dielectric barrier made of acrylic is installed on the lower electrode. The upper electrode is equipped with 10 blades shaped like knife edges. The distance between the blades is 5.3 mm, the width of each blade is 1.5 mm, and the height is 11 mm. The two end blades are 35 mm in length, and the remaining eight blades are 50 mm in length. The upper and lower electrodes are 72.5 mm in diameter with a distance between the electrodes of 7 mm, including the thickness of the dielectric barrier.

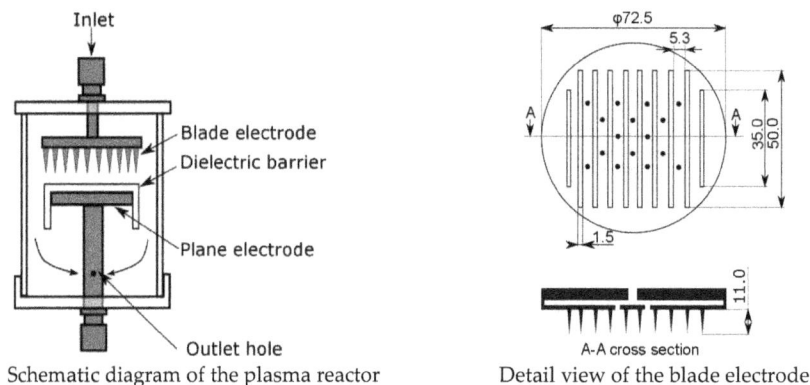

Schematic diagram of the plasma reactor

Detail view of the blade electrode

Figure 1. Schematic diagram of the plasma reactor and detail view of the blade electrode.

In this study, the practicality of the proposed NO_x removal device is considered by two methods of NO_x removal: direct oxidation and indirect oxidation. The direct oxidation method enables simulated exhaust gas to flow directly into the plasma reactor, where the nonthermal plasma directly irradiates the simulated exhaust gas, oxidizing the NO gas into NO_2. A schematic diagram of the experimental setup for the direct plasma oxidation method is shown in Figure 2. The external air is compressed with a compressor, the fine particles are removed with a filter, and the moisture is removed with a desiccant to dry the air. The dry air is mixed with NO gas supplied from a gas cylinder (2%, N_2 balance), with the flow rate and concentration regulated by mass flow controllers. After the simulated exhaust gas is passed through the plasma reactor and irradiated with the nonthermal plasma, it is passed through a heater at a flow rate of 3 L/min and heated to 270 °C to remove the O_3 that is generated in the plasma reactor. The remaining gas is then exhausted. After removal of the O_3, the simulated exhaust gas passes through the reduction process by a bubble injection-type Na$_2$S scrubber and then flows into a gas analyzer. The respective concentrations of NO, NO$_x$ (NO+NO$_2$), CO, and N$_2$O were measured with a gas analyzer (Horiba Co. Ltd., PG-240, Kyoto, Japan, chemiluminescense for NO, NO$_x$, infrared absorption for CO and O$_2$ for zirconia method,) and N$_2$O meter (Horiba Co. Ltd., VIA-510, Kyoto, Japan, infrared absorption for N$_2$O). To ensure that the pressure inside the plasma reactor remained at atmospheric pressure during the experiment, the pressure was monitored upstream of the reactor. In this experiment, the concentration of NO was set at 300 ppm, and the total flow rate flowing into the reactor was changed to 3, 5, 10, and 15 L/min.

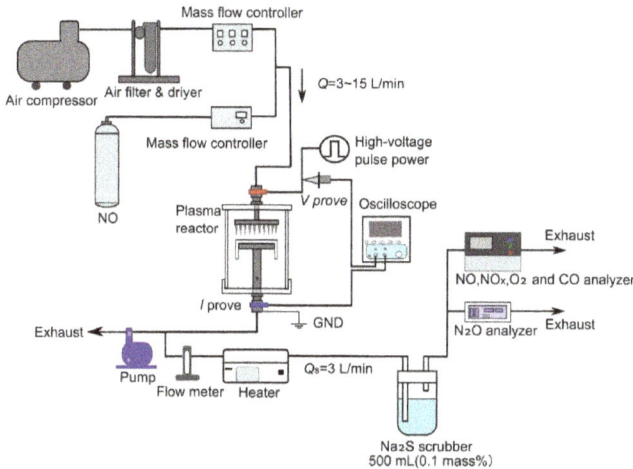

Figure 2. Schematic diagram of the experimental setup for direct plasma oxidation method.

A schematic diagram of the experimental setup for the indirect plasma oxidation method is shown in Figure 3. Unlike the direct plasma oxidation method, in which the simulated exhaust gas is directly irradiated with nonthermal plasma, in the indirect plasma oxidation method where only dry air flows through the reactor, and ozone is generated by irradiating this air with the nonthermal plasma. This ozone is then combined with the simulated exhaust gas to oxidize the NO in the gas to NO$_2$. In the direct plasma oxidation method, when the exhaust gas is at a high temperature, the gas volume expands compared to room temperature. This decreases the residence time of the gas and reduces oxidation performance. In contrast, the indirect plasma oxidation method offers the advantage that the air is treated at normal temperatures, so the residence time does not change, and oxidation performance does not decrease. The external air is compressed with a compressor, the fine particles are removed with a filter, and the moisture is removed with a desiccant to dry the air. The dry air is branched, with one part going to the plasma reactor and the other part mixed with NO gas supplied from a gas cylinder (2%, N$_2$ balance). The dry air irradiated by the nonthermal plasma in the plasma reactor is combined with the simulated exhaust gas and then passed through the heater to remove the O$_3$. The residence time of the mixture between the mixing point and the heater is 0.2–0.5 s. The residence time is enough time to react and mix NO with ozone since oxidation of NO is immediately occurred by ozone. Especially in this experiment, a tube with diameter of 4mm is used, which is quite small and considers that NO and ozone mix completely in the tube. In this experiment, the concentration of NO was set at 300 ppm and the flow rate of dry air flowing into the reactor at 1.0 L/min. The total flow rate of the air passing through the reactor and mixing with the exhaust gas was changed to 3, 5, 10, and 15 L/min.

The voltage applied to the plasma reactor was the same for the direct and indirect methods. An insulated gate bipolar transistor power supply was used to apply high-voltage pulses (max. 38 kV) at the following frequencies: 210, 420, 630, 840, and 1020 Hz. The discharge power in the plasma reactor was measured by an oscilloscope (Yokogawa Electric Corporation DL1740, Tokyo, Japan) using a high-voltage probe (Tektronix Co. Ltd., P6015A, Beaverton, USA) and an electric current probe (Tektronix Co. Ltd., P6021, Beaverton, USA). Figure 4 shows typical waveforms of the voltage, the current, and the product of the voltage and current as measured by the oscilloscope. To find the discharge power, the integral of the product of the voltage and current was calculated over one wavelength and then multiplied by the frequency. For the wet scrubber, sodium sulfide (Na$_2$S, 9-hydrate, Sigma-Aldrich, Japan Co. LLC, Tokyo, Japan) was used in an aqueous solution of 500 mL with an initial concentration of 0.1 mass%. Before processing, the pH was measured as 11.9 ± 0.2, and the oxidation-reduction potential (ORP) was measured as −300 ± 20 mV using a pH/ORP meter (Horiba Co. Ltd., D-53, Kyoto, Japan). The pH level requires careful attention because if it drops to less than 10, the Na$_2$S will react with the acid to generate highly toxic H$_2$S [18]. An H$_2$S detector (Honeywell International Inc., ToxiRAE3-H$_2$S) is therefore installed near the Na$_2$S scrubber vessel for the duration of the experiment to ensure that the pH would never drop to less than 10.

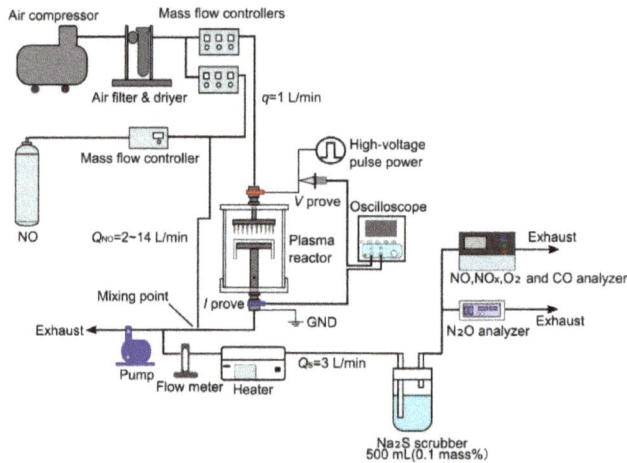

Figure 3. Schematic diagram of the experimental setup for indirect plasma oxidation method.

To evaluate the combined process of the dry plasma process and the wet chemical process using the wet scrubber, in this study NO$_x$ concentrations at various discharge power are measured. Furthermore, NO conversion efficiency and NO$_x$ removal energy efficiency are calculated and compared.

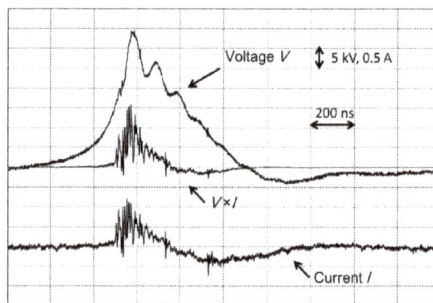

Figure 4. Waveform of voltage V, current I and instantaneous power $V \times I$ in air + NO$_x$ (300 ppm) at $Q = 5$ L/min by the direct oxidation way. (horizontal axis: 200 µs for 1 div, vertical axis: 5 kV and 0.5 A for 1 div).

4. Results and Discussion

To evaluate the NO oxidation and removal characteristics of the treatment by the combined dry plasma and wet chemical processes using the direct oxidation method, the components of the treated gas relative to the discharge power are measured. Figure 5a–d show the gas concentrations after the combined process under each flow condition (Q = 3, 5, 10, and 15 L/min). From Figure 5a–d, it is seen that the NO decreased as the discharge power increased. Furthermore, as the NO decreased, the NO_2 increased, causing a reduction in NO_x as the NO_2 is removed by the Na_2S scrubber. In Figure 5a, as the discharge power rose from its initial value to 10.5 W (V_{p-p} = 25.6 kV), the NO concentration decreased from 271 to 30 ppm, and the NO_x concentration decreased from 295 to 80 ppm. This represents an 89% reduction of NO and a 73% reduction of NO_x. In Figure 5b, when the discharge power was 9.0 W (V_{p-p} = 24.7 kV), the NO was reduced by 75% and the NO_x by 62%. In Figure 5c, when the discharge power was 11.3 W (V_{p-p} = 26.6 kV), the NO was reduced by 54% and the NO_x by 45%. In Figure 5d, when the discharge power was 10.1 W (V_{p-p} = 24.9 kV), the NO was reduced by 47% and the NO_x by 40%. These results demonstrate that treatment by the combined dry plasma and wet chemical processes effectively reduce NO_x. In Figure 5a–d, CO and N_2O were each less than 10 ppm.

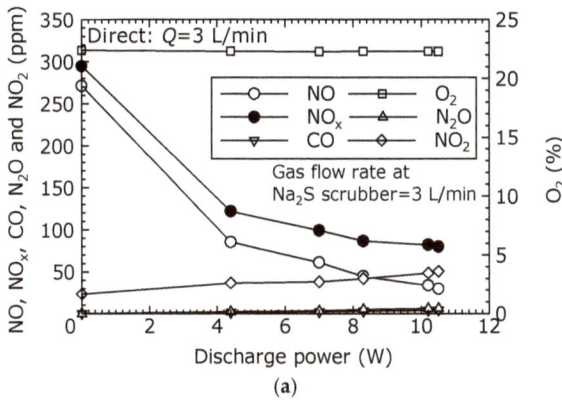

Figure 5. *Cont.*

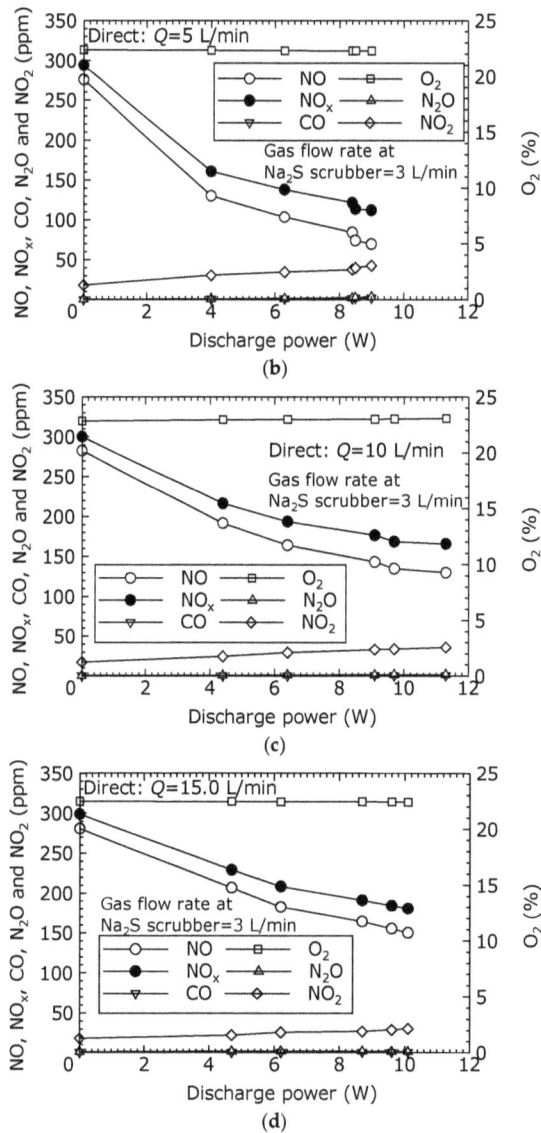

Figure 5. Gas concentrations in dry plasma and wet chemical processes by the direct oxidation method. ((a) $Q = 3$ L/min, (b) $Q = 5$ L/min, (c) $Q = 10$ L/min and (d) $Q = 15$ L/min).

The NO-to-NO$_2$ conversion characteristics of the dry plasma process are evaluated by the direct oxidation method. The relationship between the discharge power and the NO conversion efficiency at different flow rates is shown in Figure 6. In these experiments, the oxygen concentration was 22% to 24% at all flow rates. It is noted that NO$_2$ is produced at zero discharge power. It is known that a portion of NO converts to NO$_2$ upon coming into contact with oxygen by following reaction (2NO + O$_2 \rightarrow$ 2NO$_2$). Figure 6 shows that the NO conversion efficiency increases with an increase in discharge power at any flow rate. It also shows that the NO conversion efficiency increased as the flow rate decreased. As the flow rate increased, it reduced the residence time of the simulated exhaust gas in the plasma reactor, thereby decreasing the amount of oxidation promotion reactions shown in Equations (3) and (4).

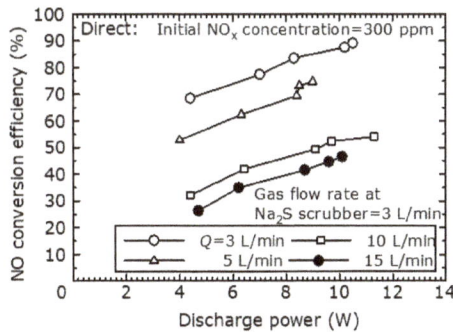

Figure 6. NO conversion efficiency varying with the discharge power at different flow rate by the direct oxidation method.

To evaluate the NO_x removal characteristics of the combined dry plasma and wet chemical processes by the direct oxidation method, the NO_x removal efficiency and NO_x removal energy efficiency at different flow rates is shown in Figure 7. The actual gas flow rate into the scrubber was fixed at 3 L/min. Here, it was assumed that the full flow rate was processed. The NO_x removal energy efficiency was calculated as the total flow rate relative to the NO_x removal amount. Figure 7 shows that as the total flow rate increased, the NO_x removal efficiency decreased, and the NO_x removal energy efficiency increased. Furthermore, at each flow rate, as the NO_x removal efficiency increased, the NO_x removal energy efficiency decreased. This is because, as shown in Figure 6, the NO oxidation efficiency rises as the discharge power increases. Within the scope of this experiment, when the NO_x removal efficiency is 60%, the NO_x removal energy efficiency is approximately 20–25 g (NO_2)/kWh.

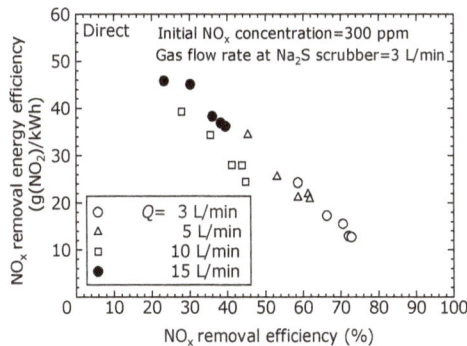

Figure 7. Measurement relation between NO_x removal efficiency and NO_x removal energy efficiency at different flow rate by the direct oxidation method. (The NO_x removal energy efficiency is calculated as total flow rate $Q = 3$–15 L/min from the inflow rate of exhaust gas to the chemical scrubber of $Q_s = 3$ L/min).

Examining Figures 6 and 7 to evaluate the efficiency of the wet chemical process, the average ratio of the decrease in NO concentration to the decrease in NO_x concentration is 89%, and the Na_2S scrubber can reduce the NO_2 over 80%. At no time was the pH of the solution after treatment greater than 11. In addition, the ORP remained at or more than −263 mV, and no H_2S was detected.

In order to evaluate the NO oxidation and removal characteristics of the combined dry plasma and wet chemical processes using the indirect oxidation method, the components of the treated gas relative to the discharge power was measured. Figure 8a–d shows the gas concentrations after the combined process under each flow condition ($q = 1$ L/min, $Q = 3, 5, 10$ and 15 L/min). From Figure 8a–d, it is seen that the NO decreased as the discharge power increased. Furthermore, as the NO decreased, the NO_2 increased, causing a reduction in NO_x as the NO_2 is removed by the Na_2S scrubber. In Figure 8a, as the discharge power rises from its initial value to 6.6 W ($V_{p-p} = 25.8$ kV), the NO concentration decreased from 281 to 7 ppm, and the NO_x concentration decreased from 298 to 52 ppm. This represents a 98% reduction of NO and an 83% reduction of NO_x. In Figure 8b, when the discharge power is 7.8 W ($V_{p-p} = 31.4$ kV), the NO is reduced by 82% and the NO_x by 66%. In Figure 8c, when the discharge power is 8.3 W ($V_{p-p} = 26.0$ kV), the NO is reduced by 43% and the NO_x by 36%. In Figure 8d, when the discharge power is 8.1 W ($V_{p-p} = 27.2$ kV), the NO is reduced by 36% and the NO_x by 32%. These results demonstrate that, using the indirect oxidation method as well, treatment by the combined dry plasma and wet chemical processes effectively reduce NO_x. In Figure 8a–d, CO and N_2O are each less than 10 ppm.

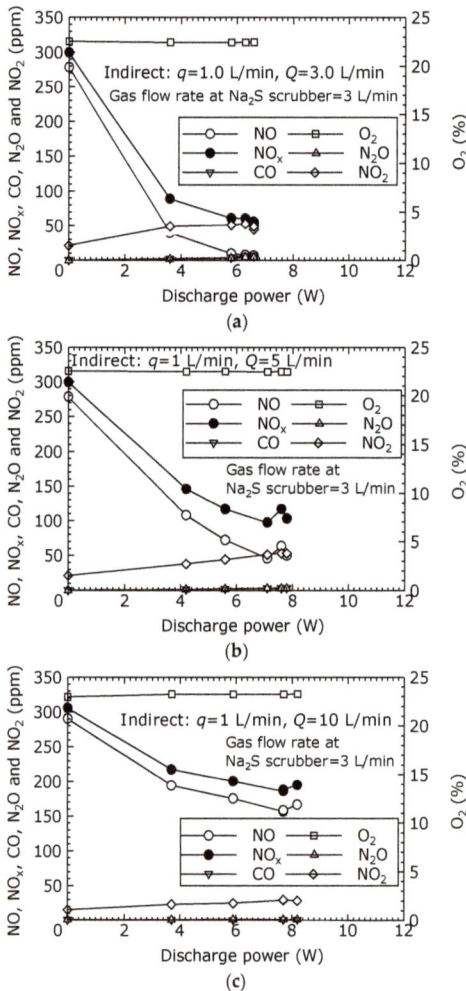

Figure 8. *Cont.*

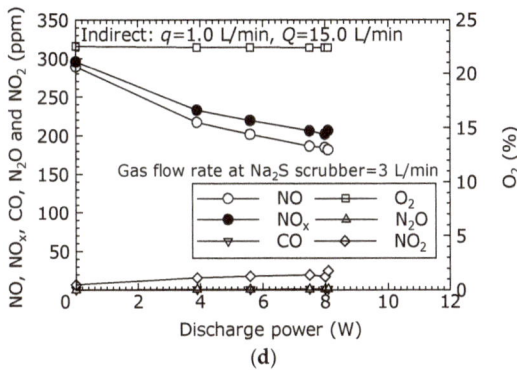

Figure 8. Gas concentrations in dry plasma and wet chemical processes by the indirect oxidation method. ((**a**) $Q = 3$ L/min, (**b**) $Q = 5$ L/min, (**c**) $Q = 10$ L/min and (**d**) $Q = 15$ L/min).

In order to evaluate the NO to NO_2 conversion characteristics of the dry plasma process by the indirect oxidation method, the relationship between the discharge power and the NO conversion efficiency at different flow rates is shown in Figure 9. In these experiments, the oxygen concentration was 22% to 24% at all flow rates. In the indirect oxidation method, even if the total flow rate is changed, the flow rate into the plasma reactor was constant ($Q = 1$ L/min), which means the ozone generated in the plasma reactor stayed constant regardless of the total flow rate. Therefore, as shown in Figure 9, the NO conversion efficiency varies by roughly the same magnitude as the flow rate (Q-q), which is the total flow rate Q minus the flow rate q flowing into the plasma reactor. However, when the total flow rate was low, the ozone might be used for further oxidation of NO_2, which suggests that an optimal flow rate may exist. In this experiment, as the flow rate increased, the NO conversion efficiency decreased, but the production of CO and N_2O was suppressed.

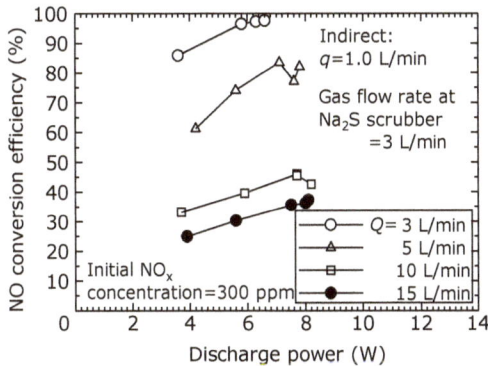

Figure 9. NO conversion efficiency varying with the discharge power at different flow rate by the indirect oxidation method.

To evaluate the NO_x removal characteristics of the dry plasma and wet chemical processes by the direct oxidation method, the NO_x removal efficiency and NO_x removal energy efficiency at different flow rates is shown in Figure 10. The NO_x removal energy efficiency was evaluated based on the molecular mass of NO_2 per discharge power with the unit of g(NO_2)/kWh. The actual gas flow rate into the scrubber was fixed at 3 L/min. Here, it is assumed that the full flow rate is processed. The NO_x removal energy efficiency was calculated as the total flow rate relative to the NO_x removal amount.

Figure 10 shows that, as the flow rate increased, the NO_x removal efficiency decreased, and the NO_x removal energy efficiency increased. As shown in Figure 10, it is clear that the flow rate affects the relationship between the NO_x removal efficiency and the NO_x removal energy efficiency. In the indirect oxidation method, the flow rate into the plasma reactor is constant at $q = 1$ L/min, which means the ozone and radical oxygen produced stay constant, regardless of the total flow rate. This explains why the energy efficiency of NO_x removal varies for each flow rate. Within the scope of this experiment, when the NO_x removal efficiency was 80%, the NO_x removal energy efficiency was approximately 22–25 g(NO_2)/kWh.

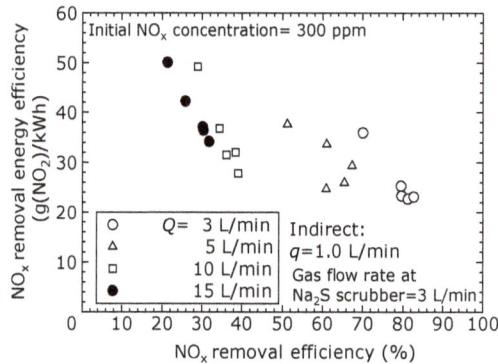

Figure 10. Measurement relation between NO_x removal efficiency and NO_x removal energy efficiency at different flow rate by the indirect oxidation method. (The NO_x removal energy efficiency is calculated as total flow rate $Q = 3$–15 L/min from the inflow rate of exhaust gas to the chemical scrubber of $Q_s = 3$ L/min).

Examining Figures 9 and 10 to evaluate the efficiency of the wet chemical process, it is observed that the average ratio of the decrease in NO concentration to the decrease in NOx concentration is 85%, and the Na_2S scrubber can reduce the NO_2 over 80%. At no time was the pH of the solution after treatment greater than 11. In addition, the ORP remained at or more than −233 mV, and no H_2S was detected.

To compare the efficiency of NO_x removal using the direct and indirect oxidation methods in this experiment, the relationship between the NO_x removal efficiency and the specific energy (SE) per unit flow rate is shown in Figure 11. The SE is calculated based on the total flow rate (3–15 L/min). The solid line is a trendline calculated by the least-squares method. Figure 11 shows no difference between the direct and indirect oxidation methods at SEs less than 50 J/L, when the NO_x removal efficiency is 20–40%. Although the indirect oxidation method generates a certain amount of ozone in the reactor regardless of the total flow rate, in this experiment, the amount of ozone generated by the indirect oxidation method is not enough to oxidize the NO in the simulated exhaust gas at the high flow rate (of 10 and 15 L/min). As the SE exceeds 50 J/L, the efficiency of NO_x removal by indirect oxidation increases, achieving an 80% removal efficiency at an SE of 132 J/L. It is considered that the amount of ozone generated by indirection oxidation method is larger than that of the direct oxidation method due to the increase of residence time as mentioned above.

These results demonstrate that the combined process of the dry plasma process using the indirect oxidation method and the wet scrubber process can remove NO_x with greater efficiency and energy savings. The plasma dry etching process used in the semiconductor industry requires a high-frequency voltage power supply and consumes several hundred to 1000 W of power. Because the power associated with the NO_x processing performed in this study was less than 10 W, the electricity consumption costs are very low. For example, assuming a 2 m^3/min exhaust class semiconductor plant that discharges NO_x at 25 g(NO_2)/h, the plasma power costs required to achieve 80% or greater NO_x removal using the proposed system at a NO_x removal energy efficiency of 25 g(NO_2)/kWh would be very economical at $0.20 /h (assuming $0.20/kWh). There is a risk that H_2S may occur as a byproduct of using Na_2S as the chemical scrubber. However, by maintaining the pH at a sufficiently high value, it is possible to reduce NO_x using Na_2S at a third of the cost of using Na_2SO_3. In the actual semiconductor manufacturing, the concentration of NO_x is emitted 100 ppm at 2000 L/min [15]. The mass flow rate is much larger than in our experiment (maximum 15 L/min). Since our experiment is fundamental research to remove NO_x at room temperature and atmospherically pressure, practical research should be performed for realizing development. The combined dry plasma and wet chemical processes can therefore be expected to serve as a low-cost and highly efficient method for treating NO_x emissions from the semiconductor manufacturing industry.

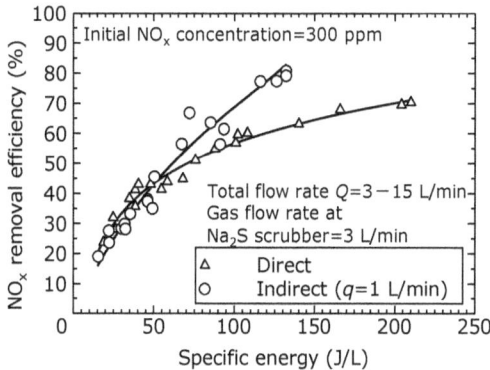

Figure 11. NO_x removal efficiency with direct and indirect oxidation method with respect to specific energy.

5. Conclusions

In this research, NO_x removal in simulate exhaust gas from semiconductor manufacturing is performed by a combined process of a direct/indirect DBD plasma process and a wet chemical process using a Na_2S scrubber. The research results contribute treatment NO_x emissions from the semiconductor manufacturing industry. We obtained the following results:

1. When the direct oxidation method was used, a maximum NO conversion rate of 89% and a NO_x removal rate of 73% at 25.6 kV and a flow rate of 3 L/min was achieved. When the indirect oxidation method was used, maximum NO conversion rate of 98% and a NO_x removal rate of 83% at 29.4 kV and a flow rate of 3 L/min was achieved. Although the N_2O and CO are detected as byproducts, they are less than 10 ppm in this experiment.
2. In the case used direct oxidation method, the NO_x removal energy efficiency was 10–25 g(NO_2)/kWh with NO_x removal efficiency of 60%. In the case used indirect oxidation method, the NOx removal energy efficiency is 20–25 g(NO_2)/kWh with NO_x removal efficiency of 80%.
3. As results comparing the NO_x removal efficiency with the specific energy, when the specific energy was more than 50 J/L, the indirect oxidation method exhibited greater removal efficiency than the direct oxidation method, achieving a maximum NO_x removal efficiency of 83% at the specific energy of 132 J/L.
4. In the NO_x reduction process by Na_2S scrubber, NO_x reduction efficiency more than 80% was achieved. In this experimental range, at no time was the pH of the solution after treatment greater than 11. In addition, no H_2S was detected.

Author Contributions: H.Y., Y.K., T.K., and M.O. performed the data processing; H.Y., T.K., and M.O. conceptualized the ideas, developed the methodology. H.Y. and Y.K. performed the experimental activity; H.Y. prepared the original draft of the manuscript; T.K. and M.O. supervised the study, reviewed the original draft.

Funding: This study is partially supported by JSPS KAKENHI Grant No. 17H03498.

Conflicts of Interest: The authors declare no conflict of interest.

References

1. Purohit, P.; Lsaksson, L.H. Global emissions of fluorinated greenhouse gases 2005–2050 with abatement potentials and costs. *Atmos. Chem. Phys.* **2017**, *17*, 2795–2816. [CrossRef]

2. Qin, L.; Han, J.; Wang, G.; Kim, H.J.; Kawaguchi, I. Highly efficient decomposition of CF_4 gases by combustion. *Sci. Res. Conf. Environ. Pollut. Public Health* **2010**, 126–130.

3. Chang, M.B.; Chang, J.S. Abatement of PFCs from semiconductor manufacturing processes by nonthermal plasma technologies: A Critical Review. *Ind. Eng. Chem. Res.* **2006**, *45*, 4101–4109. [CrossRef]

4. Lai, S.Y.; Pan, W.; Ng, C.F. Catalytic hydrolysis of dichlorodifluoromethane (CFC-12) on unpromoted and sulfate promoted TiO_2-ZrO_2 mixed oxide catalysts. *Appl. Catal. B* **2000**, *45*, 207–217. [CrossRef]

5. Takita, Y.; Morita, C.; Ninomiya, M.; Wakamatsu, H.; Ishihara, T. Catalytic decomposition of CF_4 over $AlPO_4$-based catalysts. *Chem. Lett.* **1999**, *28*, 417–418. [CrossRef]

6. Yasui, S.; Shoji, T.; Inoue, G.; Koike, K.; Takeuchi, A.; Iwasa, Y. Gas-solid reaction properties of Fluorine compounds and solid adsorbents for off-gas treatment from semiconductor facility. *Int. J. Chem. Eng.* **2012**, *2012*, 9. [CrossRef]

7. Xu, X.F.; Jeon, J.Y.; Choi, M.H.; Kim, H.Y.; Choi, W.C.; Park, Y.K. A strategy to protect Al_2O_3-based PFC decomposition catalyst from deactivation. *Chem. Lett.* **2005**, *34*, 364–365. [CrossRef]

8. Wang, Y.F.; Wang, L.C.; Shih, M.L.; Tsai, C.H. Effects of experimental parameters on NF_3 decomposition fraction in an oxygen-based RF plasma environment. *Chemosphere* **2004**, *57*, 1157–1163. [CrossRef] [PubMed]

9. Yamamoto, T.; Okubo, M.; Nagaoka, T.; Hayakawa, K. Simultaneous removal of NO_x, SO_x and CO_2 at elevated temperature using a plasma-chemical hybrid process. *IEEE Trans. Ind. Appl.* **2002**, *38*, 1168–1173. [CrossRef]

10. Fujishima, H.; Tatsumi, A.; Kuroki, T.; Tanaka, A.; Otsuka, K.; Yamamoto, T.; Okubo, M. Improvement in NO_x removal performance of the pilot-scale boiler emission control system using an indirect plasma-chemical process. *IEEE Trans. Ind. Appl.* **2010**, *46*, 1722–1729. [CrossRef]

11. Fujishima, H.; Kuroki, T.; Ito, T.; Otsuka, K.; Yamamoto, T.; Yoshida, K.; Okubo, M. Performance characteristics of pilot-scale indirect plasma and chemical system used for the removal of NO_x from boiler emission. *IEEE Trans. Ind. Appl.* **2010**, *46*, 1707–1714. [CrossRef]

12. Yamamoto, Y.; Yamamoto, H.; Takada, D.; Kuroki, T.; Fujishima, H.; Okubo, M. Simultaneous removal of NO_x and SO_x from flue gas of a glass melting furnace using a combined ozone injection and semi-dry chemical process. *Ozone Sci. Eng.* **2016**, *38*, 211–218. [CrossRef]

13. Chang, M.B.; Lee, H.M.; Wu, F.; Lai, C.R. Simultaneous removal of nitrogen oxide/nitrogen dioxide/sulfur dioxide from gas streams by combined plasma scrubbing technology. *J. Air Waste Manag. Assoc.* **2004**, *54*, 941–949. [CrossRef] [PubMed]

14. Han, B.; Kim, H.J.; Kim, Y.J. Removal of NO and SO_2 in a cylindrical water-film pulse corona discharger. *IEEE Trans. Ind. Appl.* **2005**, *51*, 679–684. [CrossRef]

15. Kim, H.J.; Han, B.; Woo, C.G.; Kim, Y.J. NO_x removal performance of a wet reduction scrubber combined with oxidation by an indirect DBD plasma for semiconductor manufacturing industries. *IEEE Trans. Ind. Appl.* **2018**, *54*, 6401–6407. [CrossRef]

16. Takaki, K.; Hatanaka, Y.; Arima, K.; Mukaigawa, S.; Fujiwara, T. Influence of electrode configuration on ozone synthesis and microdischarge property in dielectric barrier discharge reactor. *Vacuum* **2009**, *83*, 128–132. [CrossRef]

17. Malik, M.A. Nitric oxide production by high voltage electrical discharge for medical uses: A review. *Plasma Chem. Plasma Process.* **2016**, *36*, 737–766. [CrossRef]
18. Mok, Y.S. Absorption reduction technique assisted by ozone injection and sodium sulfide for NO_x removal from exhaust gas. *Chem. Eng. J.* **2006**, *118*, 63–67. [CrossRef]

energies

MDPI

Article

Development of an Electrostatic Precipitator with Porous Carbon Electrodes to Collect Carbon Particles

Yoshihiro Kawada * and Hirotaka Shimizu

Electrical Environmental Energy Engineering Unit, Polytechnic University of Japan, Tokyo 187-0035, Japan
* Correspondence: kawada@uitec.ac.jp; Tel.: +81-42-346-7765

Received: 4 June 2019; Accepted: 17 July 2019; Published: 21 July 2019

Abstract: Exhaust gases from internal combustion engines contain fine carbon particles. If a biofuel is used as the engine fuel for low-carbon emission, the exhaust gas still contains numerous carbon particles. For example, the ceramic filters currently used in automobiles with diesel engines trap these carbon particles, which are then burned during the filter regeneration process, thus releasing additional CO_2. Electrostatic precipitators are generally suitable to achieve low particle concentrations and large treatment quantities. However, low-resistivity particles, such as carbon particles, cause re-entrainment phenomena in electrostatic precipitators. In this study, we develop an electrostatic precipitator to collect fine carbon particles. Woodceramics were used for the grounded electrode in the precipitator to collect carbon particles on the carbon electrode. Woodceramics are eco-friendly materials, made from sawdust. The electrical resistivity and surface roughness of the woodceramics are varied by the firing temperature in the production process. Woodceramics electrodes feature higher resistivity and roughness as compared to stainless-steel electrodes. We evaluated the influence of woodceramics electrodes on the electric field formed by electrostatic precipitators and calculated the corresponding charge distribution. Furthermore, the particle-collection efficiency of the developed system was evaluated using an experimental apparatus.

Keywords: electrostatic precipitator; corona discharge; woodceramics; low-resistivity particle; re-entrainment phenomena; agglomeration

---

1. Introduction

In the interest of realizing a low carbon emission society, internal combustion engines powered by biofuels have been developed; however, the combustion of biofuels generates carbon particles similar to those generated from conventional fuels. Filters made from porous cordierite, silicon carbide, steel mesh are often used to collect carbon particles suspended in the exhaust gases of boilers and internal combustion engines [1]. The ceramic filters used in diesel engines called diesel particulate filters (DPFs) are regenerated by burning the collected carbon particles with the help of the diesel engine control, thus generating CO and CO_2. However, if the collected particles are gathered but not burned, the carbon density is higher than in the gas phase and the CO_2 emission can be efficiently reduced.

In an electrostatic precipitator, the particles are charged and collected on an electrode by the Coulomb force. Electrostatic precipitators offer advantages of limited pressure loss and highly effective particle collection. When the target gases have low particle concentration and high flow rates, electrostatic precipitators are suitable for particle collection [2]. Electrostatic precipitators have wide-ranging applications in, for example, thermal power plants, cement factories, and home air cleaners. The performance of an electrostatic precipitator is strongly influenced by the physical properties of the particles. The particle resistivity also affects the collection performance as particles with high resistivity remain charged for a long time. As the collected particles are charged to an inverse polarity relative to the polarity between the surface and the contacted grounded electrode, high particle resistivity results in a breakdown in the electric field across the layer of collected particles as the generated ions and charges arising from the charged particles are diminished. However, when the particle resistivity is low, charged particles are collected on the grounded electrode, and the electric charge is released; thus the collected particles are charged to the opposite polarity. The collected particles are agglomerated on the electrode surface move into the gas flow [3,4].

In general, particles in diesel exhaust feature low resistivity. In previous studies, electrostatic precipitators were attached to the diesel engines; the influence of re-entrainment phenomena was avoided by improving electrode structure and by using electrohydrodynamics [5,6], however, re-entrainment was not prevented. In another study, water or a surfactant solution was sprayed into the gas stream upstream of the electrostatic precipitator because of the addition of the liquid bridge force by the Coulomb force; this bridge force prevented the agglomerated particles on the collection electrode from moving into the gas flow. Because of the effects of the liquid bridge force with the surfactant, the agglomerated particles formed a lump-like shape rather than a pearl-chain-like structure [7]. Additionally, in another study, the polarity of the high voltage applied to the electrode were changed after a constant time interval. As a result, the collecting electrode area increased and the agglomeration particles on the electrode did not grow because of the change in the electric field direction [8]. As the collected particles remain charged, researchers have investigated the use of insulating sheets on the collector electrode [9]. In this method, the collection efficiency decreased with increasing operation time because of charging the surface of the insulating sheets.

In this study, carbon particles were collected on carbon electrodes made of woodceramics. Stainless-steel or aluminum are generally used as the electrode materials in the electrostatic precipitator. The woodceramics whose electric resistance and surface roughness were high compared with stainless-steel were used as the electrode materials. This woodceramics was formed using sawdust charcoal and phenol resin powder under specific pressure and temperature conditions; the formed pieces were then fired under a vacuum [10,11]. Woodceramics are eco-friendly materials. The electrical resistivity and pore distribution were modified by varying the firing temperature. At a firing temperature of approximately 600 °C, the wood fiber remained and the resistivity was high. When the firing temperature was above 900 °C, the carbon content was high, and the resistivity was low. The woodceramics electrodes are porous media with large surface areas on which particles may be collected. Also, woodceramics electrodes offer higher electrical resistance than stainless-steel electrodes. The electric potential near the surface of the grounded electrode might be high, thus decreasing the Coulomb force. In the present study, the electric field and charge distribution around the woodceramics electrodes are calculated and compared with those around the stainless-steel electrodes [12,13]. Furthermore, the particle collection efficiency of the proposed electrode is evaluated experimentally.

2. Calculation of the Corona Discharge Model

2.1. Calculation Model

The space charge and electrical field distributions were calculated by the finite-element method (FEM) using COMSOL Multiphysics FEM software [14,15]. The calculation model is shown in Figure 1. The air gap between the high voltage wire electrode and the grounded electrode was 9 mm the voltage supplied to the wire electrode was +8 kV, and the grounded electrode was assumed to be 0.1 mm thick in the case of stainless steel and 10 mm thick in the case of woodceramics. The electrical potential at the bottom of each electrode was 0 V. The overall mesh size was 0.4 mm or less, and the mesh above the grounded electrode was further divided of 0.08 mm in width or less. In this model, it was assumed that the woodceramics electrode was a solid material. The stainless-steel material was not defined in this model but, as a substitute a boundary condition between d and c in Figure 1 was set as 0 V. In the woodceramics, the gas-flow velocity was set to zero, other parameters of the woodceramics are shown in Table 1 [16–19]. The diffusion constant of the woodceramics was calculated using the Einstein's equation for Brownian motion as follows:

$$\frac{D}{\mu} = \frac{k \cdot T}{e},$$ (1)

where D is the diffusion constant, μ is the mobility, k is the Boltzmann constant (1.380662×10^{-23} J/K), e is the elementary charge (1.602×10^{-19} C), and T is the temperature (278 K).

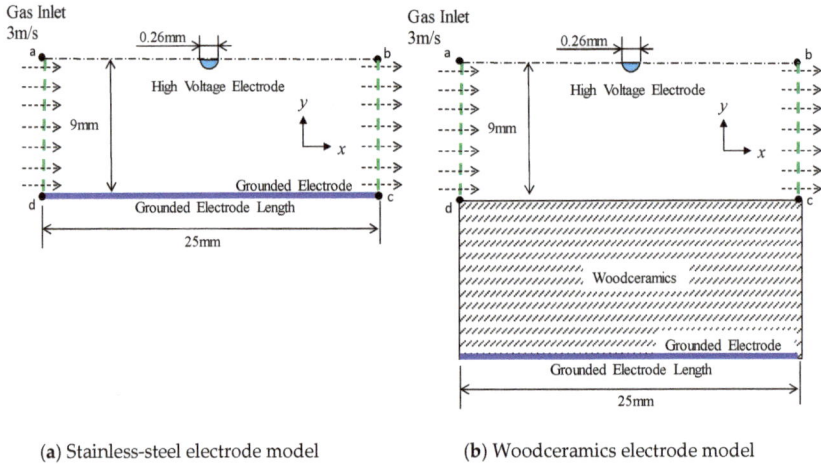

(a) Stainless-steel electrode model   (b) Woodceramics electrode model

Figure 1. Calculation models for (a) the grounded electrode with a stainless-steel electrode and (b) the grounded electrode with a woodceramics electrode (which is defined as a solid substance of 10 mm thickness).

Table 1. Calculation parameters in the air and woodceramics domains.

	In Air	Woodceramics
Mobility (m²/V·s)	2.34×10^{-4}	1.00
Diffusion constant (m²/s)	2.89×10^{-6}	0.256
Relative permittivity	1.00	5.68
Gas flow (m/s)	-	0

The space charge was analyzed by the FEM with reference to [14,15]. The spatial electric field was calculated by the following Poisson equation:

$$-\nabla \cdot \varepsilon_0 \varepsilon_r \nabla V = \rho,$$ (2)

where ε_0 is the dielectric constant, V is the electric potential, and ε_r is the relative permittivity of air (1.0). The space charge density, r, is expressed by the following equation:

$$\rho = -eN_p,$$ (3)

where N_p is the number density of positive ions (in Num/m³) and e is the elementary charge (in C). The applied voltage and number density of charges were multiplied by tanh (10^5 t) (where t is the time in seconds) to obtain 90% of the applied voltage at 15 μs and 99.5% of the number density at 30 μs. The charge density N_p is given around the wire. However, the calculations revealed that there was a gap between high density and zero density in the unmodified model. Hence, given the central position of the wire was $x = 0$, N_p was multiplied by the following parameter:

$$\alpha = 1 - \frac{x^2}{l^2},\tag{4}$$

where x is the position in the horizontal direction, and l is half of the wire radius. It should be noted that this calculation method does not result in attenuation compared with the real charge density. The number density of positive ions, N_p, was calculated as follows:

$$N_p = \frac{I_d}{L_w} \times \frac{1}{\mu_p \times L_g} \times \frac{L_{Air}}{V_{ap}} \times \frac{1}{e} \times \frac{1}{2},\tag{5}$$

where I_d is the corona discharge current, V_{ap} is the applied voltage, L_w is the length of each wire electrode (88 mm × 2), and μ_p is the mobility of the positive ions (2.34 × 10^{-4} m^2/V·s).

The transport equation for positive ions, including the ion wind and the gas flow, is given in Equation (6).

$$\frac{\partial N_p}{\partial t} + \nabla \cdot \left(-D_p \nabla N_p - \mu_p E N_p\right) + \nabla \cdot \left(N_p U_g\right) = 0,\tag{6}$$

where D_p is the diffusion constant of the positive ions in the air (2.89 × 10^{-6} m^2/s), U_g is the gas flow (in m/s), and E is the electric field strength (in V/m):

$$E = -\nabla V.\tag{7}$$

Note that this equation does not account for thermal diffusion.

The Navier–Stokes equation is used to model the gas flow and ion wind as follows:

$$\rho_g \frac{\partial U_g}{\partial t} + \rho_g \left(U_g \cdot \nabla\right) U_g = -\nabla P + \mu_g \nabla^2 U_g + F,\tag{8}$$

where P is the pressure (in Pa), r_g is the air density (1.205 kg/m^3), μ_g is the dynamic coefficient of air viscosity (1.822 × 10^{-5} Pa·s), and F is the external Coulomb force (in N), which is the product of the space charge density, ρ and the electric field E.

2.2. Numerical Results

The potential distributions are shown in Figure 2. The contour lines represent the equipotential line from 500 to 4000 V. With the woodceramics electrode, the potential is high near the grounded electrode. In addition, the ion distributions over 2.0 × 10^{14} Num/m^3 are shown in Figure 3. The two graphs are similar in shape demonstrating that the ion distribution area in the corona discharge has not changed. Due to the influence of the electrode material, the electric field and ion number distribution were evaluated directly below the wire electrode. The electric fields between the wire and the grounded electrode are shown in Figure 4a. The electric field at the surface of the woodceramics electrode was approximately 10% less than that at the surface of the stainless-steel electrode. The ion distribution between the wire and the grounded electrode is shown in Figure 4b. This curve was smoothed with a moving average algorithm over an averaging distance of 0.48 mm for approximately 20 sampling points. With the stainless-steel electrode, the ion density slightly fluctuated. The ion density increased close to the electrodes but remained comparable for the two types of electrodes.

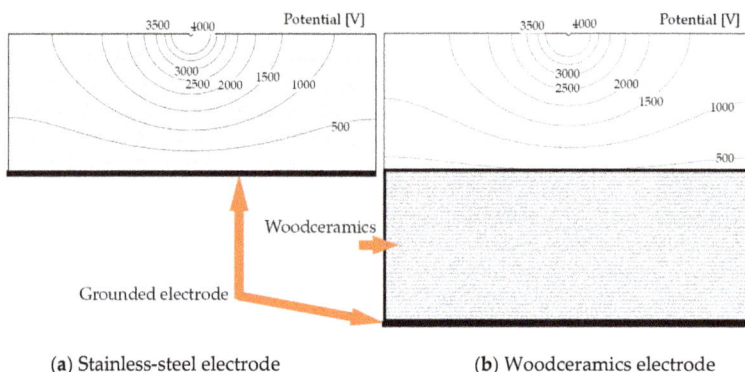

(**a**) Stainless-steel electrode   (**b**) Woodceramics electrode

Figure 2. Electrical potential of the corona electrode structure with the (**a**) stainless-steel grounded electrode and (**b**) woodceramics grounded electrode. The bottom side of the woodceramics was grounded, so, the potential near the woodceramics surface was approximately 480 V because of the current flow in the woodceramics.

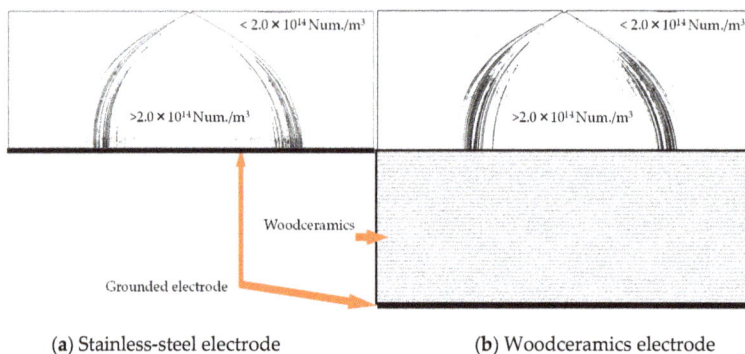

(**a**) Stainless-steel electrode   (**b**) Woodceramics electrode

Figure 3. Illustration of the charge number densities of charges at the (**a**) stainless-steel grounded electrode and (**b**) the woodceramics electrode. These distributions are similar in shape to the charge distribution. It can be seen that the resistivity of the grounded electrode does not have strong influence on the overall charge distribution.

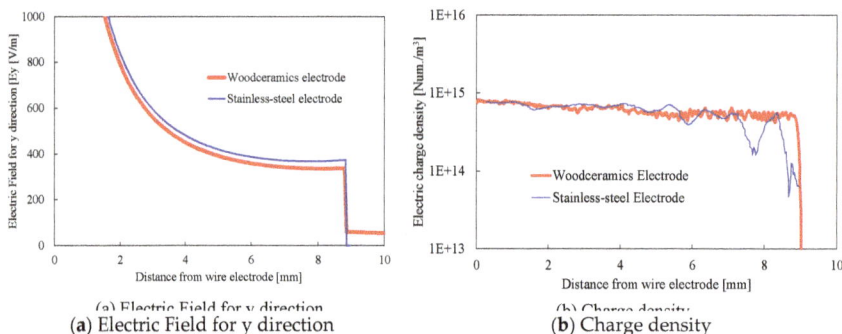

(**a**) Electric Field for y direction   (**b**) Charge density

Figure 4. (**a**) Electric field in the y direction and (**b**) charge density between the wire electrode and grounded electrode. The horizontal axes represent the distance from the wire electrode. With the woodceramics electrode, the electric field decreased by 10% and the charge density was slightly lower than that with the stainless-steel electrode.

These results shed light on the particle-collection process; the particle charge and immigration velocity decrease when using a woodceramics electrode, unlike with a stainless-steel electrode, because of the decrease in the electric field.

3. Experimental Methods

3.1. Woodceramics

Figure 5 shows the process of manufacturing woodceramics by molding charcoal powder or a part of a plywood board. In this study, sawdust charcoal was used as the charcoal powder; it was pulverized to a particle size of less than 1 mm and mixed with phenol resin powder (Kanebo, Bellpearl S899) at a weight ratio of 8:2 using a ball mill. The mixed powder was placed in a mold temporarily while degassing at 200 °C and 4.0 MPa. Thereafter, the samples were compressed at 160 °C and 20 MPa and cooled from 160 °C to 40 °C for 10 min. The samples were then molded elsewhere. Under a vacuum of less than 50 Pa, the molded samples were fired at 300 °C or 600 °C for 3 h to complete the woodceramics manufacturing process. Microscopic photographs were taken using a digital still microscope (KEYENCE, VH-5000) (see Figure 6). In the woodceramics, the wood fibers remained on the surface, which was rougher than that of the stainless-steel electrode.

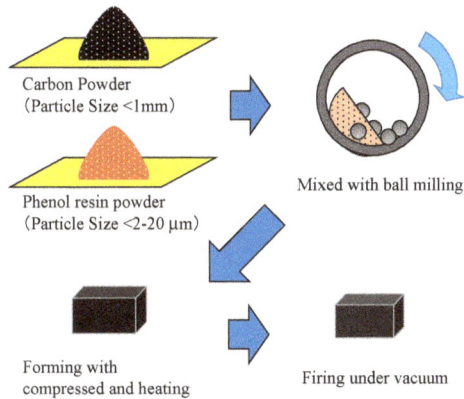

Carbon Powder
(Particle Size <1mm)

Phenol resin powder
(Particle Size <2-20 μm)

Mixed with ball milling

Forming with
compressed and heating

Firing under vacuum

Figure 5. The process of manufacturing woodceramics. First, the carbon powder was made by mixing sawdust charcoal and phenol resin powder by ball milling. Then, the mixed powder was compressed and heated. Finally, the formed test piece was fired under a vacuum.

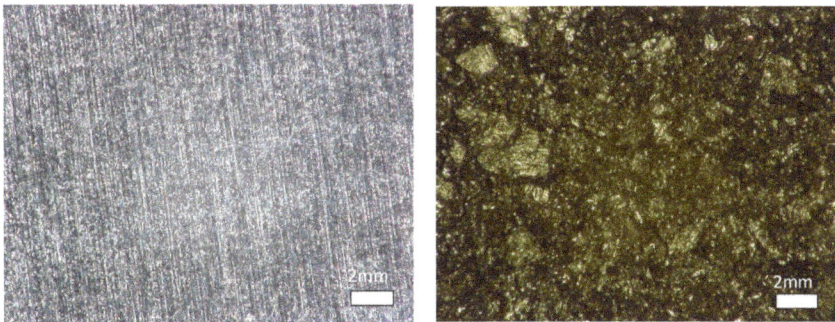

(**a**) Stainless-steel plate electrode (**b**) Woodceramics plate electrode

Figure 6. (**a**) Stainless-steel surface and (**b**) woodceramic surface. On the stainless-steel surface, the marks of the grinder are seen. On the woodceramic surface, the wood fiber appears dotted.

The electric resistivities of the woodceramics and stainless-steel electrodes were measured according to the Japanese Industrial Standard (JIS K7194 Testing method for resistivity of conductive plastics with a four-point probe array) using the measurement probe is shown in Figure 7. The needles were sewing needles with radii of curvature of $R = 70$ μm and diameters of $\Phi = 0.71$ mm. Woodceramics and stainless-steel electrodes were prepared for each of three samples, and five points were measured on each sample. A current of 1 or 10 mA was flowed between A and D while the voltage between B and C was measured with a source measure unit (Yokogawa GS610). The contacts at the tips of the needles and the test piece were adjusted to a minimum voltage between B and C. The resistivity was calculated as follows:

$$\rho = F \cdot t \cdot R, \tag{9}$$

where r is the resistivity (in Ω·cm), t is the thickness (in cm) of the test piece, and R is the resistance (in Ω) obtained from the current and voltage measurements. The thickness, t was in the range of 0.106–0.138 mm for the stainless-steel electrodes and in the range of 11.4–11.9 mm for the woodceramics electrodes. Thus, the correction factor for the thickness, F was 4.2353 for the stainless-steel electrodes and in the range of 2.3693–2.4014 for the woodceramics electrodes.

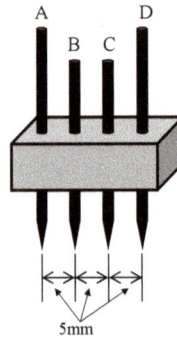

Figure 7. Four terminal probes used for measuring resistivity. Terminals A to D are needles of the same shape. Current flowed from A to D or from D to A, while the voltage was measured between B and C.

The average resistivities are shown In Table 2. The resistivity of the woodceramics material was 98.2 Ω·cm, which was 4400 times higher than that of stainless steel which was measured as 0.0281 Ω·cm.

Table 2. Resistivities of woodceramics and stainless-steel.

	Resistivity (Ω·cm)	
	Stainless-Steel	Woodceramics
Average	0.0218	96.2
Max.	0.032	137
Min.	0.0143	67.3

3.2. Experimental Setup and Conditions

A schematic diagram of the experimental apparatus is shown in Figure 8. The test particles were generated by the combustion of vegetable oil. Particle-laden gas was introduced via the experimental duct with the updraft, and the gas was diffused by a diffusion fan. Then, the sample gas was absorbed into the electrostatic precipitator and treated with a positive corona discharge and diffused by the fan. The treated gas then resided in the downstream test duct. The number density of particles in this gas was measured with a particle counter (RION, KC-01E + diluter KD-01); the number densities of the room air and target gas are shown in Table 3. The concentration of 0.3–0.5 μm particles in the target gas was 1.2×10^5 Num/L and over 80% of these particles originated from the combustion. To verify the particle concentration, the sampling position was moved to upstream of the electrostatic precipitator before and after each test was completed. Thus the particle-collection efficiency was calculated as follows:

$$\eta = \left(1 - \frac{N_{out}}{N_{in}}\right) \times 100 \ (\%) \ , \tag{10}$$

where N_{in} indicates the upstream particle number concentration, and N_{out} is the downstream particle number concentration of the electrostatic precipitator. Instead of N_{in}, the averaged particle concentration measured downstream of the electrostatic precipitator without an applied voltage was used.

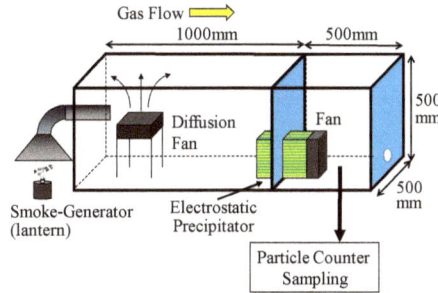

Figure 8. Schematic diagram of the experimental apparatus. An oil lantern was used as the particle generator. The particle concentration was measured downstream of the electrostatic precipitator.

Table 3. Average concentrations of particles in various size ranges in the room air and target gas.

Particle Size Range (μm)	Room Air (Num/L)	Target Gas (Num/L)
0.3–0.5	1.6×10^4	1.2×10^5
0.5–1.0	9.6×10^2	5.8×10^3
1.0–2.0	3.0×10	5.8×10^2
2.0–5.0	0	1.7×10^2

3.3. Electrostatic Precipitator

A top-down view of the electrostatic precipitator is shown in Figure 9. The gas treated with corona discharge passed through the strainer with the perforated metal plate and was channeled to the downstream exhaust duct. The gas-flow velocity was adjusted by varying the fan speed. The fan and the gas flow were stopped when measuring the discharge current. The mean gas velocity was kept at 3 m/s when evaluating the particle collection.

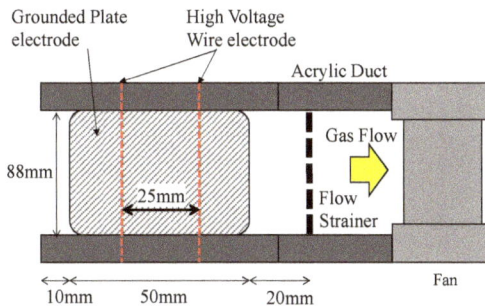

Figure 9. Top view of the electrostatic precipitator. The gas flow is from the left to the right. The treated gas propelled the fan, and the exhaust gas flows into the downstream duct.

The electrode structures of the corona discharge with the woodceramics grounded electrode and stainless-steel grounded electrode are shown in Figure 10. The two grounded electrodes were placed parallel to each other at a distance of 18 mm apart, and the two wire electrodes for the applied high voltage were set at the midpoint between the grounded electrodes. The electrostatic precipitator duct was 88 mm in total width and 18 mm high. The wire electrode was connected to a DC + high-voltage power supply (Matsusada Precision, HAR-20R15), and the discharge current was measured with an ammeter attached to the power supply. In each grounded electrode, the distance between the wire electrode and grounded electrode was 9 mm. In addition, the thickness of the woodceramics was approximately 10 mm, and a copper grounded electrode was attached behind the woodceramics.

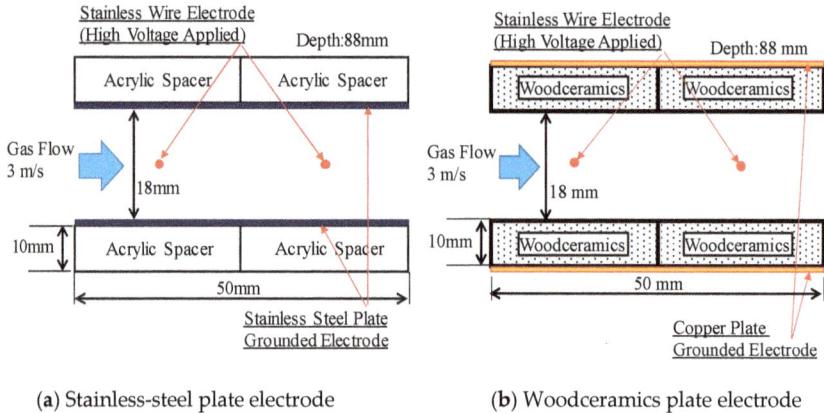

(a) Stainless-steel plate electrode (b) Woodceramics plate electrode

Figure 10. Side view of each electrode structure. In the corona discharge electrode configuration, the grounded electrode was either a (a) stainless-steel grounded electrode or (b) woodceramics grounded electrode. Each grounded plate electrode structure was placed such that its surface was the same distance from the wire electrode.

4. Results and Discussion

Figure 11 shows the discharge current as a function of the applied voltage. The corona on-set voltage was approximately 6.0 kV. The current-voltage curves for the woodceramics and stainless-steel grounded electrodes were very similar.

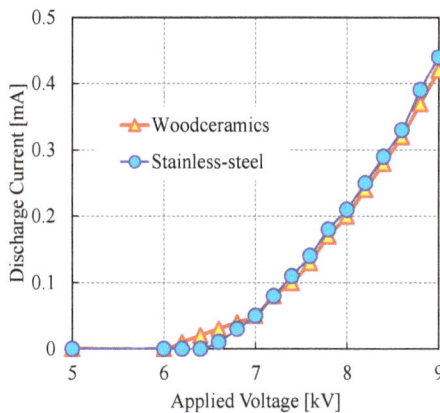

Figure 11. Corona discharge current as a function of the applied voltage for various electrodes. The discharge current characteristics of the two electrodes follow similar trends.

Figure 12 shows the collection efficiencies using woodceramics and stainless-steel grounded electrodes for operating times of 0–5 min, 5–10 min, 10–15 min, and 15–20 min. The particle-collection efficiency with the woodceramics electrode was high and remained constant over the entire operating period. On the other hand, when using the stainless-steel electrode, the particle-collection efficiency was constant for particles smaller than 1 μm during the operating time, but that for particles larger than 1 μm decreased over the operating period. This decrease can be attributed to the particle agglomeration on the electrode surface and the occurrence of particle re-entrainment phenomena. However, such particle re-entrainment phenomena were prevented by using the woodceramics electrode. Thus, the surface roughness and the electrical resistivity of woodceramics improve the collection efficiency. In future studies, we will investigate the mechanism underlying the agglomeration of the collected particles and the effect of improving the particle collection using a woodceramics electrode.

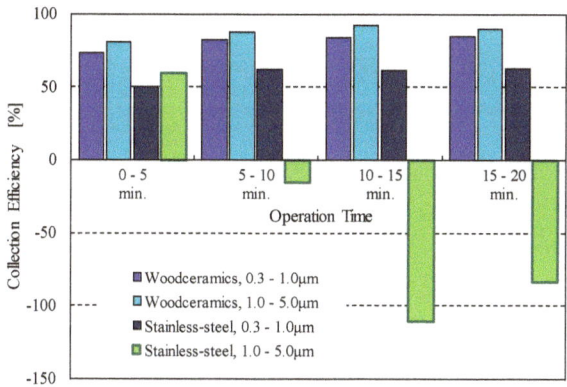

Figure 12. Particle-collection efficiencies over time for various electrode types and particle size ranges. With the woodceramics electrode, the collection efficiency for large particles remained high while that with the stainless-steel electrode decreased with the operating time.

5. Conclusions

Stainless-steel or aluminum are generally used as electrode materials in the electrostatic precipitator. In this study, we utilized a novel electrostatic precipitator to collect carbon particles on a carbon electrode based on charged particles and the Coulomb force. A woodceramics made from sawdust charcoal which has an electrical resistivity 4400 times higher than stainless steel was used as the carbon electrode. Because the woodceramics material has a porous surface, it was expected to improve the particle collection. The electric field and charge distribution were calculated and the actual particle collection was evaluated experimentally.

The calculation results showed that, using woodceramics electrodes, the electric field near the electrode surface was reduced by approximately 10%, and the particle-collection efficiency was reduced accordingly. However, the adhesion force of the particles after contact with the electrode depended on the intermolecular force and not the Coulomb force. Therefore, the only concern was the reduction in the migration speed.

Our experimental results showed that, with the use of the woodceramics electrode, the particle-collection efficiency improved in all particle size ranges while the re-entrainment phenomenon decreased. However, the present study serves only as a feasibility study. In future research, the applied voltage characteristics and flow velocity characteristics should be further tested.

Author Contributions: Conceptualization, Y.K. and H.S.; Methodology, Y.K.; Software, Y.K.; Validation, Y.K.; Formal Analysis, Y.K.; Investigation, Y.K. and H.S.; Resources, Y.K. and H.S.; Data Curation, Y.K.; Writing-Original Draft Preparation, Y.K.; Writing-Review & Editing, Y.K.; Visualization, Y.K.; Supervision, H.S.; Project Administration, Y.K.; Funding Acquisition, Y.K. and H.S.

Energies **2019**, *12*, 2805

Funding: This research received no external funding.

Acknowledgments: The authors would like to thank Enago for the English language review.

Conflicts of Interest: The funders had no role in the design of the study; in the collection, analyses, or interpretation of data; in the writing of the manuscript, and in the decision to publish the results.

References

1. Pajdowski, P.; Puchałka, B. The Process of Diesel Particulate Filter Regeneration under Real Driving Conditions. *IOP Conf. Ser. Earth Environ. Sci.* **2019**, *214*, 012114. [CrossRef]
2. Godish, T. *Indoor Air Pollution Control*, 3rd ed.; Lewis Publishers: Chelsea, MI, USA, 1991; pp. 247–263.
3. White, H.J. *Industrial Electrostatic Precipitation*; Addison-Wesley Publishing: Reading, MA, USA, 1963; pp. 319–329.
4. Takahashi, T.; Zukeran, A.; Ehara, Y.; Ito, T.; Takamatsu, T.; Kawakami, H. Influence of Re-entrainment Phenomena on Particle Deposit in Electrostatic Precipitator. *IEEJ Trans. Fundam. Mater.* **1999**, *119*, 254–260. [CrossRef]
5. Yamamoto, T.; Abe, T.; Mimura, T.; Otsuka, N.; Ito, Y.; Ehara, Y.; Zukeran, A. Electrohydrodynamically Assisted Electrostatic Precipitator for the Collection of Low-Resistivity Dust. *IEEE Trans. Ind. Appl.* **2009**, *45*, 2178–2184. [CrossRef]
6. Masuda, S.; Moon, J.-D. Electrostatic Precipitation of Carbon Soot from Diesel Engine Exhaust. *IEEE Trans. Ind. Appl.* **1983**, *19*, 1104–1111. [CrossRef]
7. Kawada, Y.; Jindai, W.; Zukeran, A.; Ehara, Y.; Ito, T.; Kawakami, H.; Takahashi, T. Influence of Lyphophile on Preventing Re-entrainment in Spraying Surfactant Type Electrostatic Precipitator. In Proceedings of the 7th International Conference of Electrostatic Precipitation, Kyonju, Korea, 20–25 September 1998.
8. Kubo, T.; Kawada, Y.; Takahashi, T.; Ehara, Y.; Ito, T.; Zukeran, A.; Takamatsu, T. The relation between shape of particles and collection efficiency by electrostatic precipitators. *J. Aerosol Sci.* **2000**, *31*, 452–453. [CrossRef]
9. Takahashi, T.; Kawada, Y.; Zukeran, A.; Ehara, Y.; Ito, T. Inhibitory effect of coating electrode with dielectric sheets on re-entrainment in electrostatic precipitator. *J. Aerosol Sci.* **1998**, *29*, S485–S486. [CrossRef]
10. Shibata, K.; Kasai, K.; Okabe, T.; Saito, K. Development of Porous Carbon Material "Woodceramics"—Electrical Properties with in Low-Temperature Region. *J. Soc. Sci. Jpn.* **1995**, *44*, 284–287. (In Japanese) [CrossRef]
11. Kawada, Y.; Shimizu, H. C-H Bonds and Phenol Resin Contents in Woodceramics under Fabrication. In Proceedings of the 25th Annual Meeting of MRS-J, Yokohama, Japan, 8–10 December 2015.
12. Kawada, Y.; Shimizu, H.; Ohkawa, M.; Mori, S.; Kakishita, K. Elect of Woodceramics Grounded Electrode on Electrostatic Precipitation with Positive Corona Discharge. *Trans. Mater. Res. Soc. Jpn.* **2018**, *43*, 187–190. [CrossRef]
13. Kawada, Y.; Shimizu, H.; Ohkawa, M.; Mori, S.; Kakishita, K. A Study on Woodceramics Collector in Electrostatic Precipitator. In Proceedings of the Annual Meeting of IESJ, Osaka, Japan, 11–12 September 2017; pp. 27–30. (In Japanese).
14. Takeuchi, N. Simulation of Ionic Wind Induced by Corona Discharge Using COMSOL Multiphysics. *J. Inst. Electrost. Jpn.* **2016**, *40*, 168–171. (In Japanese)
15. Kawada, Y.; Shimizu, H.; Zukeran, A. Numerical Study of the Suitable Precharger Grounded Electrode Length in Two-Stage-Type Electrostatic Precipitators. *IEEE Trans. Ind. Appl.* **2019**, *55*, 833–839. [CrossRef]
16. The Institute of Electrostatics Japan (Ed.) *Handbook of Electrostatics*; Ohmsha, Ltd.: Tokyo, Japan, 1980. (In Japanese)
17. Chang, J.S.; Kelly, A.J.; Crowley, J.M. (Eds.) *Handbook of Electrostatic Processes*; Marcel Dekker, Inc.: New York, NY, USA, 1995.
18. Waker, P.L., Jr.; Thrower, P.A. (Eds.) *Chemistry and Physics of Carbon, Volume 16*; Marcel Dekker, Inc.: New York, NY, USA, 1981; p. 141.
19. Okabe, T. (Ed.) *Woodceramics*; Uchida Rokakuho Publishing: Tokyo, Japan, 1996; p. 65. (In Japanese)

energies

MDPI

Article

Fundamental Evaluation of Thermal Switch Based on Ionic Wind

Keiichiro Yoshida

Department of Electric and Electronic Systems Engineering, Osaka Institute of Engineering, Osaka 535-8585, Japan; keiichiro.yoshida@oit.ac.jp; Tel.: +81-6-6954-4236

Received: 20 June 2019; Accepted: 30 July 2019; Published: 1 August 2019

Abstract: A significant amount of thermal energy (mainly under 200 °C) is wasted in the world. To utilize the waste heat, efficient heat management and thermal switching is required. The basic characteristics of a thermal switch that controls the flow of heat by switching on/off the ionic wind is discussed in this study. The study was conducted through experiments and numerical simulations. A heater made of aluminum block maintained at 100 °C was used as a heat source, and the rate of heat flow to a copper plate placed over it was measured. Ionic wind was induced by corona discharge with a needle placed on the heater. The ratio of heat transfer coefficients was obtained in the range of 3–4, with an energy efficiency of around 10. The heat flux at this condition was approximately 400 W/m^2. The numerical simulations indicate that the heat transfer is enhanced by ionic winds, and the results were found to corroborate well with the experimental ones.

Keywords: thermal switch; ionic wind; corona discharge; thermal management; waste heat

1. Introduction

The global thermal energy waste amounts to a significant fraction of the total energy consumption. For instance, only the industrial sector in Japan is discharging around 3.3×10^{18} J/y heat as waste [1], despite the enormous efforts that have been put for waste heat recovery since the oil crisis in 1973. In fact, the amount corresponds to approximately 40% of the total energy consumption in Japan [2], indicating that efficient waste heat recovery is crucial to save fossil fuels and reduce CO_2 emission.

Therefore, a thermal switch that controls the flow of heat is essential. Literature shows that a thermal switch could stabilize the high temperature at the hot end of a thermoelectric element and improve the electrical output [3]. Further, the low energy conversion efficiency of thermoelectric devices limit their use in a low temperature range (<200 °C, waste heat in this range contributes to 60% of the waste heat from the Japanese industry [1]), and heat energy has to be utilized as heat if possible. However, a thermal switch enables the thermoelectric device to operate in low temperature ranges.

Various thermal switch technologies have been studied in the literature including mechanical control of the contact between two surfaces [3], the phase transition of VO_2 between electrical conductor and insulator [4], the use of liquid–vapor phase transition of working fluid [5], and electrowetting of liquid [6–9]. Nonetheless, the devices are desired to be simple without any moving parts and should be able to control the flow of heat at any temperature to have precise and long-term thermal management for an effective heat recovery. The reported techniques fail to meet all of these requirements in one. Thus, the present author proposes a simple method that uses ionic wind. In this method, the ionic wind induced by corona discharge between two surfaces controls convective heat transfer. The device has a simple structure and can control heat flow at any temperature without any moving part.

It is worth mentioning that heat transfer enhancement by using ionic wind devices has been studied for decades [10–19]. For example, [18] showed that ionic wind induced by a wire-type discharge electrode enhanced the heat transfer between two of the surfaces that face each other of a rectangular duct (the duct had a primary airstream). Reference [19] showed numerical simulation results that represent the heat transferred by the ionic wind induced by a wire-type corona electrode from heat sources attached on the outer surface of a circular duct to the air inside the duct. However, as for now, those studies have been simply aimed at the enhancement of cooling objects and lack either of the following two requirements as the thermal switch base on ionic wind: (1) the device has both heat emitter and heat receptor surfaces, and (2) heat flow should be as small as possible when corona wind is off (i.e., it has no primary airstream). In this study, a device that met the above-mentioned conditions was prepared, and the basic characteristics of a thermal switch that uses ionic wind were investigated. Particularly, (a) the ratio of heat transfer coefficient between the surfaces in the presence and absence of corona discharge, and (b) the ratio of controlled rate of heat flow to the power consumed by corona discharge were evaluated through experiments and numerical simulations.

2. Materials and Methods

2.1. Experimental Methods

Figure 1 shows the schematic of the setup for thermal switch tests. The temperature of the 10 mm thick and 50 mm square aluminum block was maintained at 100 °C using an embedded nichrome-wire heater and a temperature controller. This configuration supposed a minimum unit that is taken out of a thermal switch which has multiple discharging electrodes in a regular pattern. The electric current and driving voltage of the heater were obtained using voltages V_1 and V_2 (refer to Figure 1). A 0.5 mm thick copper plate was placed at a vertical distance of 12.5 mm from the block. The heat transfer happened between upper surfaces of the aluminum block and the copper plate. Corona needles with 1 mm diameters and different heights (L = 6–9 mm) were placed at the center on the block. The needles were made from brass and polished with an abrasive compound before use. When high positive DC voltage, V_c, was applied to the copper plate with respect to the aluminum block (connected to the ground), negative ions were generated at the vicinity of the needle tip and ionic wind was induced in the gap. The electric current due to the corona discharge was calculated by measuring the voltage across a 10 kΩ resistor (refer to Figure 1).

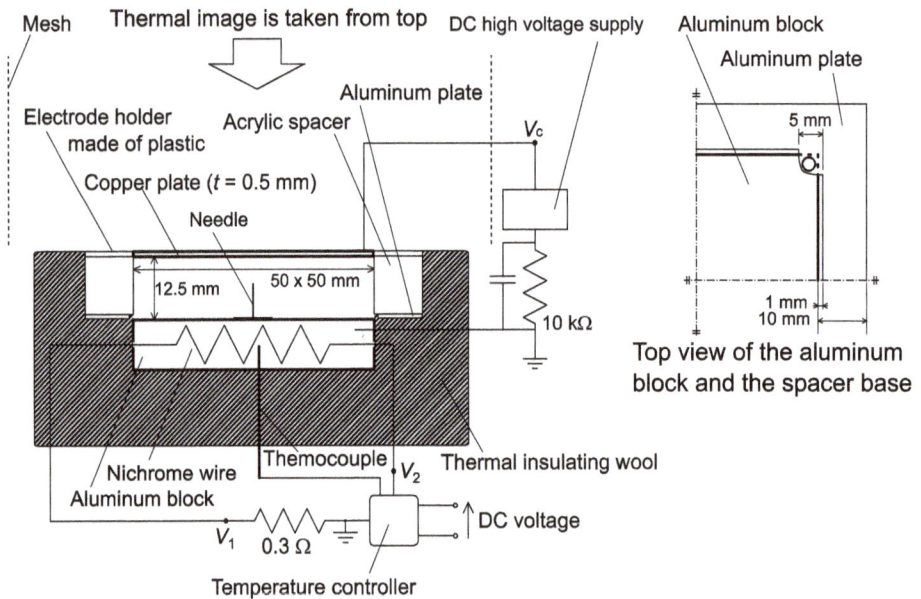

Figure 1. Schematics of the thermal switch test setup.

The copper plate and the aluminum block were separated with acrylic spacers. To minimize unintended heat conduction (a) the acrylic spacers were placed on an aluminum plate ($t = 1$ mm) that was connected only at four corners of the square, and (b) 1 mm gaps were provided between the edges of the block and the spacer unit (shown in detailed illustration presented in Figure 1). The set was thermally insulated by using ceramic wool. Because the ceramic wool is permeable to air, the gap space was not strictly sealed, but probably air slightly entered and exited through the 1 mm gap. The complete setup was enclosed in a plastic mesh to restrict any wind from the surrounding without affecting the natural convection over the copper plate.

The temperature at the top surface of the copper plate was measured using a thermographic camera. This temperature could not be measured with contacting probes such as thermocouples because of the high voltage applied to the copper plate. The top surface of the copper plate and the electrode holder was painted with a black paint (suspension of graphite particles, F-142 manufactured by Fine Chemicals Japan Co., Ltd., Tokyo, Japan) to avoid the measurement errors caused by emissivity less than one. Although the Infrared camera can observe the temperature distribution, the standard deviations of the temperature over the copper surface were approximately found to be 0.3 °C. The heat transfer coefficient (K) is defined as,

$$K = \frac{Q}{(T_h - T_s)A},\tag{1}$$

where Q is the rate of heat flow (in Watt) from the surface of the aluminum block to the copper plate, A is the area of the surfaces between which heat is transferred (equal to the surface area of the aluminum block and the copper plate in the present case), T_h is the temperature of the top surface of the aluminum block, and T_s is the average temperature of the top surface of the copper plate. Q was determined by the power supplied to the heater to maintain T_h at 100 °C. Further details to determine Q are discussed in the results section.

2.2. Numerical Simulation

Airflow and transfer of heat induced by corona discharge were numerically simulated using finite element method in COMSOL Multiphysics 5.3. The calculations were performed on an axisymmetric region having 25 mm radius, as illustrated in Figure 2. The region was vertically divided into two parts—one part had air (12.5 mm), and the other had copper (0.5 mm). A needle was placed in the air gap such that its axis aligned with the axis of the region and its base touched the bottom surface. The needle consisted of a conical tip (with 2 mm height and 0.05 mm radius of curvature of the tip) and a cylindrical base of 0.5 mm radius. The radius of curvature of the tip was determined from the needle typically used in such experiments (Figure 3). Henceforth, "the tip" indicates the surface that has a radius of curvature of 0.05 mm. The sizes of the elements (for finite element method) at the surface of the needle tip were kept less than 2 μm × 2 μm to calculate ionic flow correctly.

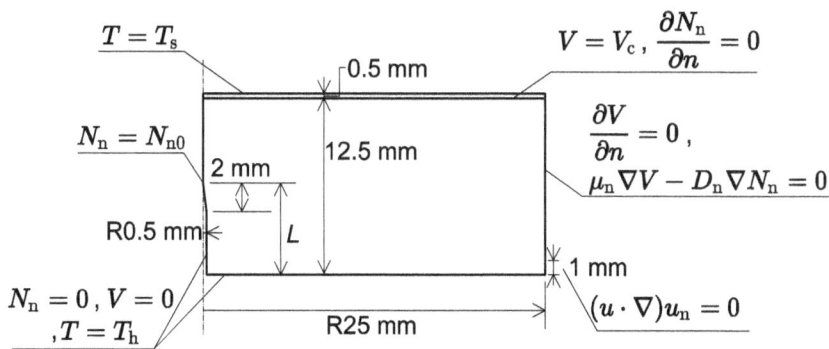

Figure 2. Calculation region and boundary conditions.

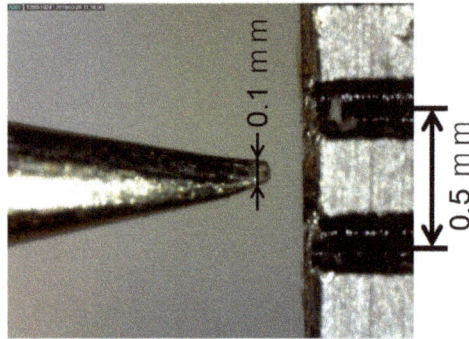

Figure 3. The tip of a corona needle.

To simulate the corona discharge, electrical potential (V) and negative ion density (N_n) in the air-filled region were estimated using Poisson's Equation (2) and continuity Equation (3). Further, the space charge was assumed to consist of negative ions of molecular oxygen (O_2^-) only. Due to two orders of magnitude higher speed of ions than airstream in corona discharge (in general), the effect of airflow on the transportation of ions was neglected.

$$\nabla \cdot (\varepsilon_0 \nabla V) = eN_n \tag{2}$$

$$\nabla \cdot (\mu_n \nabla V - D_n \nabla N_n) = 0 \tag{3}$$

Here e is the elementary charge, μ_n is the mobility of the negative ions, and D_n is the diffusion coefficient of the negative ions. Mobility μ_n was calculated by multiplying $T_g/293.15$ with the value at 293.15 K (1.69×10^{-4} m^2/V·s) [20,21], where T_g (=$(T_h + T_s)/2$) is the average temperature in the air-filled region. Further, the Einstein's relation was used to determine D_n using μ_n at temperature T_g.

The simulations were performed under the following boundary conditions. The boundary conditions for the potential (V) were determined using the Dirichlet conditions (Equations (4) and (5)) at the top surface and the bottom surface including the needle and the Neumann condition (Equation (6)) at the sidewall. Likewise, Neumann boundary condition (Equation (7)) was used for ion density N_n at the top surface of the air-filled region. The ion flux was set to zero at the sidewall (Equation (8)). Further, the ion density at the bottom surface except for the tip of the needle was set to zero (Equation (9)), while the ion density at the tip was set to N_{n0} (Equation (10)). The following procedure was followed to determine N_{n0}: i) first, the corona onset voltage (V_o) was determined for each needle length (L) through experiments, ii) second, the maximum electric field strength (E_{max}) at the tip of the needle (the field strength when V_o is applied to the needle without the corona discharge) was calculated, and iii) third, N_{n0} was adjusted so that the maximum field strength at the tip became E_{max} when V_c was applied to the needle. This procedure is based on Katpzov's hypothesis [22] and validated through numerical simulation, e.g., by Umezu et al. [23].

$$V = V_c \tag{4}$$

$$V = 0 \tag{5}$$

$$\frac{\partial V}{\partial n} = 0 \tag{6}$$

$$\frac{\partial N_n}{\partial n} = 0 \tag{7}$$

$$\mu_n \nabla V - D_n \nabla N_n = 0 \tag{8}$$

$$N_n = 0 \tag{9}$$

$$N_n = N_{n0} \tag{10}$$

The velocity field (u) of the laminar flow induced by the corona discharge was determined by the Navier–Stokes Equation (11) at a stationary state. The equation uses volume force (eN_nVV) exerted by the ion movement and buoyancy force ($-\rho_a g$) to determine u. To solve the equation, the wall was considered to have zero slip (Equation (12)). However, the thin opening between the acrylic spacer and the aluminum block (see the experimental setup section) was represented by setting stress normal to the bottom 1 mm part of the sidewall to zero (Equation (13)).

$$\rho_a(u \cdot \nabla u) = -\nabla p + \eta \Delta u - \rho_a g + eN_n \nabla V \tag{11}$$

$$u = 0 \tag{12}$$

$$(u \cdot \nabla)u_n = 0 \tag{13}$$

Here ρ_a is the density of air, p is a static pressure, η is the viscosity of air, and g is the acceleration due to gravity. Air density and viscosity are the functions of temperature and were determined as discussed below.

The temperature field (T) in the air-filled region and at the copper plate was calculated to determine the rate of heat flux. Dissipation Equations (14) and (15) govern the temperature field in the air-filled and the copper-filled regions, respectively. Boundary condition (16) was used at the surface of the heater block and the needle. The experimentally measured value of T_s was used in boundary condition (17). The temperature gradient in the normal direction at the bottom 1 mm opening of the sidewall was set to zero (Equation (18)).

$$\rho_a C_{p,a} u \cdot \nabla T + \nabla \cdot (-k_a \nabla T) = 0 \tag{14}$$

$$\nabla \cdot (-k_c \nabla T) = 0 \tag{15}$$

$$T = T_h \tag{16}$$

$$T = T_s \tag{17}$$

$$\frac{\partial T}{\partial n} = 0 \tag{18}$$

In the above equations, $C_{p,a}$ is the isobaric specific heat of air, k_a and k_c are the thermal conductivity of air and copper, respectively. k_a approximated by Equation (19) (Reference [24]) at T_g (= ($T_s + T_h$)/2) was used.

$$k_a = 7.46 \times 10^{-5}(T - 273.15) + 2.48 \times 10^{-2}(W/(mK)) \tag{19}$$

3. Results and Discussion

3.1. The Heat Transfer by Natural Convection and Analysis of the Rate of Heat Flow (Q)

In the absence of a needle in the gap, the heat is transferred by natural convection from the heater surface to the copper plate. This section discusses the rate of heat flow (Q) by natural convection. Figure 4 shows the variation of heater powers with respect to elapsed time in two cases: (a) when the copper plate is exposed to the atmosphere, as illustrated in Figure 1, and (b) when the copper plate is insulated from the atmosphere by using ceramic wool. Since a significant amount of heat leaks from the sidewalls and the bottom surface of the heater, the difference in the heater powers in two cases is expected to be equivalent to the rate of heat flow from heater to copper plate by natural convection. The average difference in the heater powers in two cases is 0.36 W (the stabilized heater powers are 6.17 and 5.81 W). The average temperature of the top surface of the copper plate (T_s) is 46.8, while the temperature of the heater (T_h) is 101 °C. With this temperature difference, only the thermal conduction provides approximately 0.33 W heat flow. That implies most of the heat is transferred by thermal conduction alone.

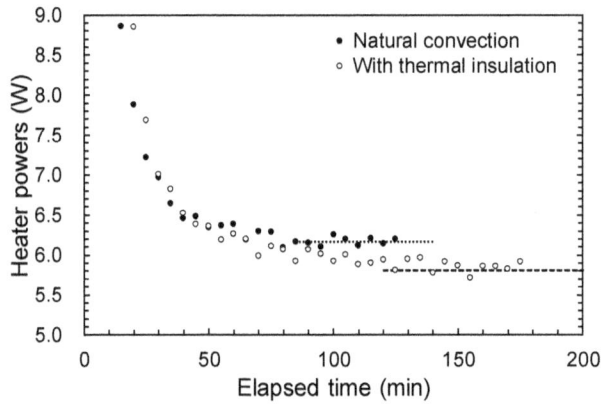

Figure 4. Heater powers in the cases with and without thermal insulation over the copper plate. (No corona discharge was applied in the gap).

Figure 5 shows the result of numerical simulation in the absence of ionic wind in the gap. The temperature gradient is almost uniform, with no influence of the opening at the bottom of the sidewall. The result supports the claim that heat was transferred mainly by conduction in the absence of ionic wind. The simulated value of Q (0.258 W) is less than the experimental value because the surface area of the heater is smaller in numerical simulation.

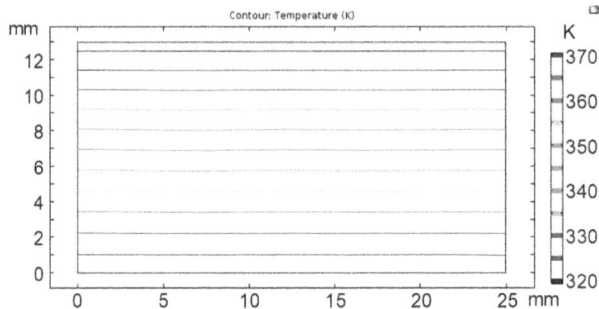

Figure 5. Temperature distribution in the gap when only natural convection occurs (numerical simulation).

In the purview of the above results, herein, Q in the absence of ionic wind can be approximated as 0.33 W. Accordingly, the heat flow rate in the presence of ionic wind can be given by

$$Q = P_{h1} - P_{h0} + 0.33(W), \tag{20}$$

where P_{h0} is the heater power upon stabilization before the corona discharge, and P_{h1} is that 30 min after the beginning of the corona discharge.

3.2. The Effect of Ionic Wind

3.2.1. The Corona Onset Voltages (V_o)

Figure 6 shows the experimental results of corona discharge current (I_c) as a function of applied DC voltage (V_c) for needle length L ranging from 6 mm to 9 mm. In most of the cases, Equation (21) with a constant C [25] very closely approximates I_c (V_o in the equation is the corona onset voltage). However, for $L = 9$ mm some outliers against the fitting curve at lower voltages (less than 3 kV) can be observed, probably because of oxidation of the needle tip.

$$I_c = CV_c(V_c - V_o) \tag{21}$$

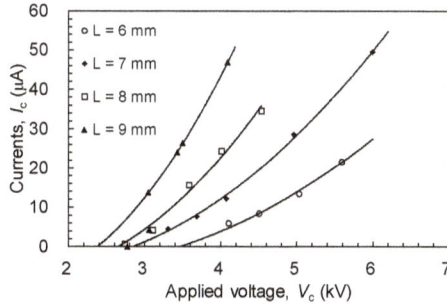

Figure 6. Discharge currents (I_c) vs. applied voltage (V_c) (experiment).

Table 1 summarizes V_o as obtained by fitting I_c vs. V_c data using Equation (21). It is observed that the onset voltage decreases with decrease in the gap between the needle tip and the copper plate.

Table 1. Corona onset voltage (V_o) vs. the length of corona needle (L).

L (mm)	V_o (kV)
6.0	3.46
7.0	2.85
8.0	2.65
9.0	2.38

Figure 7 shows I_c as a function of V_c, as obtained from numerical simulations. Like the experimental results, the relationship between V_c and I_c can be expressed by Equation (21), and I_c is larger for larger L. However, the magnitudes of the corona discharge currents are significantly smaller than that in the experiment. Interestingly, the difference between the simulated and the experimental I_c values increases with L, e. g., the numerically simulated values are around one-fifth of the experimental values for $L = 6$ mm, while the same are one-tenth of the experimental values for $L = 9$ mm. It may be due to high local electric field around the nonspherical tip of the needle causing easier breakdown of air; however, the exact reason is not known at this stage.

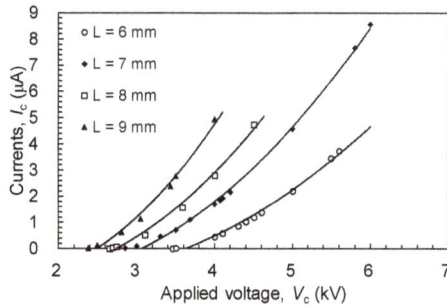

Figure 7. Applied voltage (V_c) vs. discharge currents (I_c) (numerical simulation).

3.2.2. The Relationship Between Applied Voltage and the Rate of Heat Transfer

Figure 8a–d show the experimental and simulated (a) rate of heat flow (Q) normalized by the heat transfer area (A), and (b) temperature difference between the heater and the top surface of the copper plate ($T_h - T_s$) for each L. The area (A) used in experiments and simulations are 2.5×10^{-3} m^2 and 1.963×10^{-3} m^2, respectively. For every L, the average heat flux, Q/A, increases with increase in V_c, while the surface temperature of the copper plate T_s approaches the heater temperature T_h. The numerical simulations reproduced the experimental trend between the applied voltage and Q/A. The following discussion further justifies the accuracy of the simulated airstream. In the experiment, an increase in Q causes a decrease in the temperature difference between T_h and T_s, which in turn decreases Q (negative feedback for Q works). Here, T_s obtained from the experiment was used as a boundary condition in the simulation. Accordingly, if the simulated convection were (a) less active than that in the experiment, Q would not have increased with V_c, and (b) more active than that in the experiment, Q would have increased more rapidly with V_c than was observed in the experiment. The best agreement between experimental and simulated results was observed for $L = 7$ mm, which could be attributed to the status of the tip of the needle (although any deviation was not apparent). In addition, a few calculations were performed for the V_c that were not tested in the experiment. In such cases, the boundary value T_s was linearly interpolated from the experimental values.

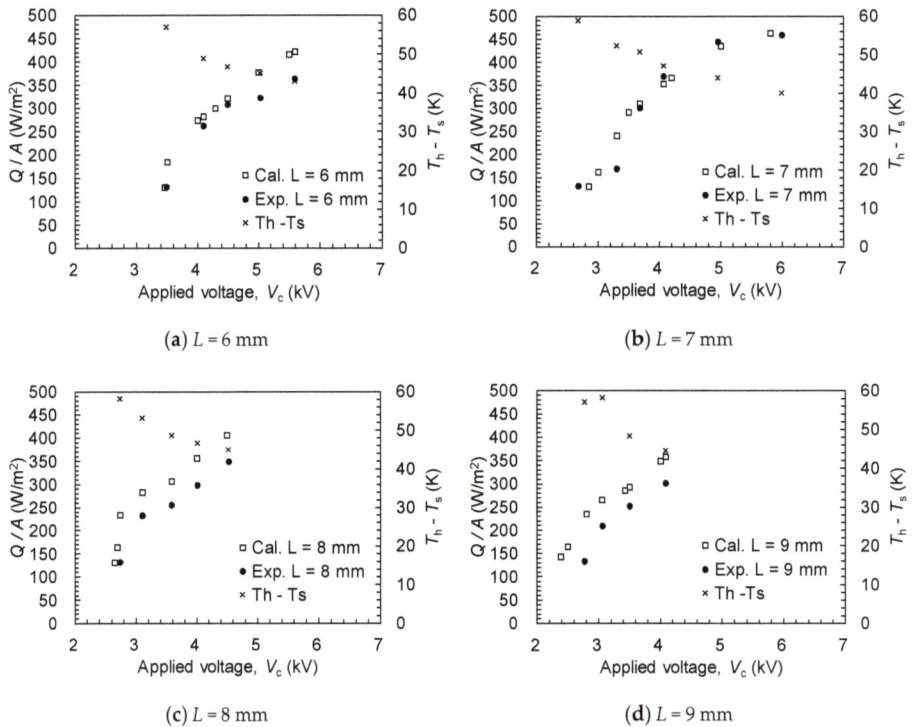

Figure 8. (a–d) Average heat flux (Q/A) and temperature difference ($T_h - T_s$).

Table 2 summarizes the heat transfer coefficient K. The ratio K_{on}/K_{off} defines the ability of the device to control the flow of heat. K_{on}/K_{off} can be increased by applying higher voltages. The maximum value for the same in the present study is 5.12 ($V_c = 6.0$ kV and $L = 7$ mm). At 4.0–4.1 kV, K increases with L (experimentally as well as in the simulation), except for $L = 7$ mm. The result is an outcome of the smaller gap between the needle tip and copper plate resulting in higher electric field and larger discharge current. However, the optimum L in the context of energy efficiency is discussed in the next section.

Table 2. Heat transfer coefficients (K) and the ratio of K between with and without corona discharge (K_{on}/K_{off}).

V_c (kV)	K (W/(K·m²))		K_{on}/K_{off}	
	Experiment	Calculation	Experiment	Calculation
		$L = 6$ mm		
0	2.32	2.31	-	-
3.5	2.32	3.25	1	1.41
4.1	5.36	5.77	2.31	2.50
4.5	6.58	6.88	2.84	2.98
5.0	7.16	8.41	3.08	3.64
		$L = 7$ mm		
0	2.32	2.31	-	-
3.3	3.24	4.60	1.45	2.06
3.7	5.94	6.14	2.65	2.75
4.1	7.85	7.49	3.50	3.35
5.0	10.1	9.90	4.50	4.43
6.0	11.5	11.7	5.12	5.25
		$L = 8$ mm		
0	2.32	2.31	-	-
2.7	2.32	4.01	1.0	1.78
3.1	4.38	5.32	1.93	2.36
3.6	5.26	6.30	2.32	2.79
4.0	6.39	7.62	2.82	3.37
4.5	7.74	9.02	3.41	4.00
		$L = 9$ mm		
0	2.32	2.31	-	-
2.8	2.33	4.11	1.01	1.79
3.1	3.59	5.42	1.55	2.36
3.5	5.23	6.06	2.26	2.63
4.1	6.79	8.06	2.93	3.50

Figure 9a–f shows the simulated distributions of temperature and air flow velocity for $L = 6$ mm. It is evident that an increase in the speed of flow with the applied voltage increases the temperature gradient on the copper plate right above the needle and on the peripheral area of the heater; hence, enhances the rate of heat transfer. Three types of ionic wind patterns appeared depending on the applied voltages. At the lowest voltage (3.5 kV), when the airstream above the needle reached the copper plate it turned outward, descended, cooled down, reached the sidewall, and then returned to the axis. Thus, the wind circulated clockwise. Regardless, a slight clockwise circulation occurred at a peripheral region due to natural convection. At V_c = 4.0 kV, the circular flow at the center region induced a counterclockwise circulation at the peripheral region. At the highest V (5.0 kV), the discharge-induced flow at the axis went outward along the copper plate and reached the sidewall, then descended before being cooled. It resulted in the formation of a high-speed flow layer near the center of the upper surface which grows with increase in V_c. Further, the flow layer breaks down the temperature boundary layer and results in an enhanced rate of heat transfer. Additionally, for the other values of L, the same orders of flow pattern were observed including a single circulation localized to the central region, twin circulation, and a global single circulation, as it was for $L = 6$ mm.

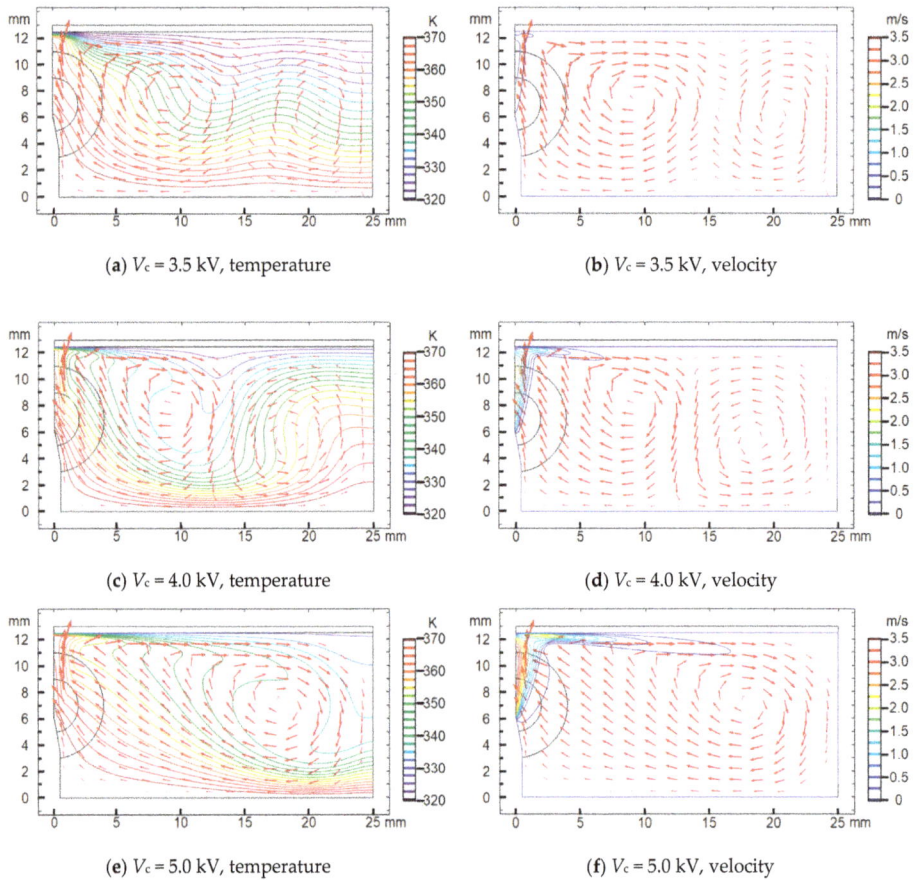

(a) V_c = 3.5 kV, temperature

(b) V_c = 3.5 kV, velocity

(c) V_c = 4.0 kV, temperature

(d) V_c = 4.0 kV, velocity

(e) V_c = 5.0 kV, temperature

(f) V_c = 5.0 kV, velocity

Figure 9. (**a**–**f**) Temperature and velocity distributions in the gap (numerical simulation, L = 6 mm; arrows are velocity vector indicating relative strength of the flow with logarithmic scale).

Figure 10 shows the Nusselt number (Nu) as a function of the Reynolds number (Re) based on the numerical simulation results. The gap height d = 12.5 mm is adopted as the representative length d. The representative velocity is the average airstream speed in the radial direction at 1 mm below the upper surface of the gap ($|u|_{ave}$ defined by Equation (23)). There is a significant correlation between Nu and Re. It should be noted that this correlation does not depend on L and this correlation means that the effect of the change in the convection regime (twin or single circulation) can be reduced to Re. It can be understood that the flow in the radial direction induces the circulating flows in the gap, which transfer heat.

$$Nu = \frac{Kd}{k_a} \qquad (22)$$

$$|u|_{ave} = \int_0^{25mm} 2\pi r |u| dr \qquad (23)$$

$$Re = \frac{|u|_{ave} d}{\mu} \qquad (24)$$

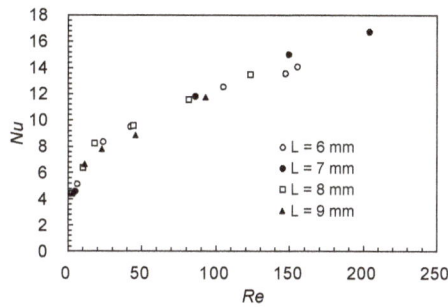

Figure 10. The Nusselt number as a function of the Reynolds number based on the numerical simulation.

3.2.3. Energy Efficiencies

In the present study, energy efficiency (η) can be defined as the ratio of the increased rate of heat flow (ΔQ) to the input discharge power P_c ($= I_c \times V_c$) (see Equation (25)). The experimental and simulated values of the same are shown in Figures 11 and 12, respectively.

$$\eta = \Delta Q / P_c, \tag{25}$$

where $\Delta Q = Q_{on} - Q_{off}$ (Q_{on} is Q in the presence of corona discharge, and Q_{off} is Q in the absence of corona discharge). As evident from Figure 11, the efficiency is extremely high if Q/A is less than 200 W/m^2, then it decreases rapidly with increase in the voltage. As can be found in Figure 10, in the almost absence of convection, only a small radial flow has a large enhancement effect for heat transfer, which is the reason for high η at low Q/A region. The decrease in η with higher Q/A is attributed to decrease in conversion efficiency from electrical energy to the kinetic energy of airstream. In the range of Q/A larger than 250 W/m^2, the needle with $L = 7$ mm resulted in the highest η, followed by 6, 8, and 9 mm (the order is hard to determine in the range less than 250 W/m^2). The reason of the low efficiencies for larger values of L is the less amount of air in the space above the needles, which reduces the effect of the discharge on the rate of heat transfer. Further, $L = 6$ mm did not give the highest efficiency because of the weaker outward flow compared with that of $L = 7$ mm that resulted due to the lower velocity at the axis. However, the reason of larger variation in η in the experiment than in the numerical simulation is not fully understood in this study. Among the reasons would be that the incomplete spherical shape of the needle tips might have increased current that did not contribute to creation of the air flow, especially in cases of $L = 8$ and 9 mm (see Section 3.2.1). The energy efficiency is more than one for all the conditions in the range of this study. For example, it is ~10 when L is 7 mm and Q/A is ~400 W/m^2 ($K_{on}/K_{off} = 3$–4). In a viewpoint of energy consumption, these results mean that transferring thermal energy from where waste heat is exhausted has an advantage compared with newly creating heat at the site. The simulations' results (Figure 12) show the same trend as the experiment, with the highest η for $L = 7$ mm. However, the much larger efficiencies obtained from simulations are due to smaller discharge current compared with that of the experiment. Analysis indicates that for $V_c = 4.1$ kV and $L = 7$ mm, Q/A and η are 370 W/m^2 and 12.0, respectively, in the experiment, while the corresponding simulated values are 353 W/m^2 and 58.0. Very large values of efficiency obtained from simulation show a great potential of the ionic wind-based thermal switches.

Although this device is promising from the energy efficiency point of view, the heat flux needs to be 2–3 orders of magnitude higher for practical purposes. Accordingly, while keeping in mind the advantage of ionic wind that it can be generated in a small region (like the heat transfer gap in this study), the future studies should identify the optimum arrangement of multiple needles that could generate mutually driving winds, causing a very strong circulating flow.

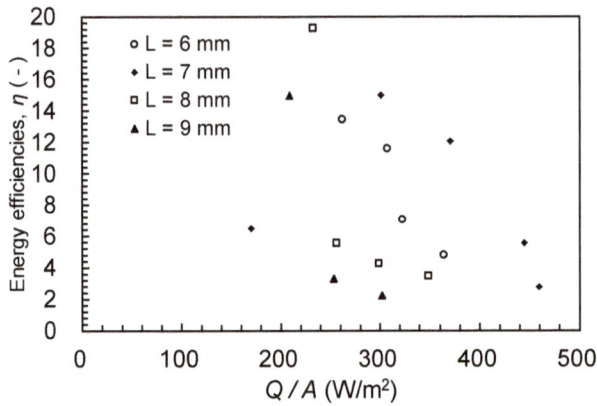

Figure 11. Energy efficiency vs. heat transfer rate (experiment).

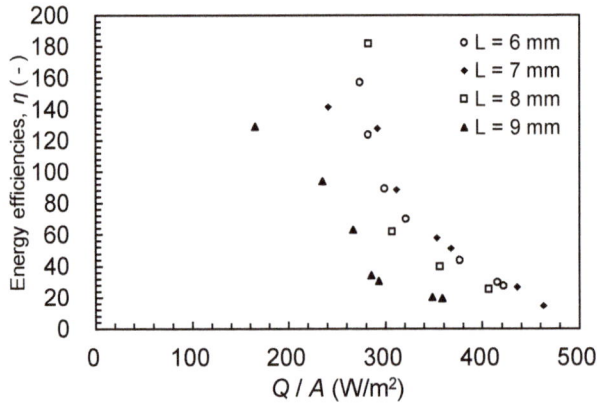

Figure 12. Energy efficiency vs. heat transfer rate (numerical simulation).

3.2.4. The Influence of the Heat Generation by Corona Discharge in the Gap

The numerical simulation results discussed until now did not include the heating effect of the corona discharge. All of the energy from the corona discharge was dissipated in the gap. Since it is difficult to identify the distribution of such a heat source, heat transferred was calculated by assuming a volume heat source distributed uniformly throughout the gap. Heating power per unit volume was determined using Equation (21) at various applied voltages for the case of $L = 6$ mm. It should be noted that experimental values of the discharge power were used in the calculation. The simulated values of the rate of heat transfer after considering the volume heat sources are shown in Figure 13. Due to a little temperature rise in the gap, the rate of heat flow from the heater surface decreased, while the heat received by the upper surface of the gap increased. The difference (between black triangles and circles in the graph) in the rates of heat flow correspond to the discharge power. The introduction of a volume heat source has less effect on the rate of heat flow at the heater surface as compared with the upper surface (copper plate). This result suggests that the experimental values of Q are hardly affected by the discharge power dissipated in the gap because Q is determined as the heater power in the experiment.

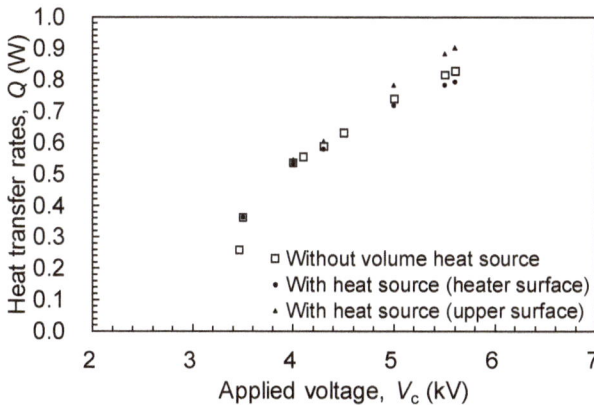

Figure 13. Heat transfer rates with and without the discharge power dissipated in the gap (numerical simulation using the discharge powers that were determined experimentally).

4. Conclusions

To summarize, the basic characteristics of the thermal switch using ionic wind were investigated in this study. The ionic wind-based thermal switches could control the rate of heat flow, with energy efficiencies (the ratio of controlled heat flow rate to input discharge power) more than 10. Further, the ratio of heat transfer coefficients between two surfaces in the presence and absence of corona discharge ranged from 3 to 4, and the heat flux was around 400 W/m^2. The future study should focus to enhance the heat flux in a controlled manner.

Funding: This research received no external funding.

Acknowledgments: The author thanks undergraduate students, Yoshiki Chayama, Ryoya Moriguchi, and Seiichiro Kitano, very much for their help in conducting the experiments.

Conflicts of Interest: The author declares no conflict of interest.

References

1. Kato, Y.; Suzuki, H.; Shikazono, N. Heat storage, transportation, and transfer. In *Energy Technology Roadmaps of Japan: Future Energy Systems Based on Feasible Technologies Beyond 2030*; Kato, Y., Koyama, M., Eds.; Springer: Cham, Switzerland, 2016; pp. 135–146.
2. *Key World Energy Statics*; International Energy Agency: Paris, France, 2018; Available online: https://www.iea.org/ (accessed on 1 June 2019).
3. Gou, X.; Ping, H.; Ou, Q.; Xiao, H.; Qing, S. A novel thermoelectric generation system with thermal switch. *Appl. Energy* **2015**, *160*, 843–852. [CrossRef]
4. Gu, W.; Tang, G.-H.; Tao, W.-Q. Thermal switch and thermal rectification enabled by near-field radiative heat transfer between three slabs. *Int. J. Heat Mass Transf.* **2015**, *82*, 429–434. [CrossRef]
5. Velson, N.V.; Tarau, C.; Anderson, W.G. Two-Phase Thermal Switch for Spacecraft Passive Thermal Management. In Proceedings of the 45th International Conference on Environmental Systems, Bellevue, WA, USA, 12–16 July 2015.
6. Cha, G.; Kim, C.-J.; Ju, Y.S. Thermal conductance switching based on the actuation of liquid droplets through the electrowetting on dielectric (EWOD) phenomenon. *Appl. Therm. Eng.* **2016**, *98*, 189–195. [CrossRef]
7. Gong, J.; Cha, G.; Ju, Y.S.; Kim, C.-J. Thermal Switches Based on Coplanar EWOD for Satellite Thermal Control. In Proceedings of the 2008 IEEE 21st International Conference on Micro Electro Mechanical Systems (MEMS), Wuhan, China, 13–17 January 2008.
8. McLanahan, A.L.R.; Richards, C.D.; Richards, R.F. A Dielectric Liquid Contact Thermal Switch with Electrowetting Actuation. In Proceedings of the ASME 2010 International Mechanical Engineering Congress & Exposition IMECE2010, Vancouver, BC, Canada, 12–18 November 2010; pp. 61–66.

9. Tasaki, Y. Present status and approach to a room-temperature magnetic refrigerator with thermal switches. *J. Cryog. Supercond. Soc. Jpn.* **2015**, *50*, 53–59. [CrossRef]
10. Velkoff, H.R.; Godfrey, R. Low-velocity heat transfer to a flat plate in the presence of a corona discharge in air. *J. Heat Transf.* **1979**, *101*, 157–163. [CrossRef]
11. Franke, M.E.; Hutson, K.E. Effect of corona discharge on free-convection heat transfer inside a vertical hollow cylinder. *J. Heat Transf.* **1984**, *106*, 347–351. [CrossRef]
12. Tada, Y.; Takimoto, A.; Hayashi, Y. Heat transfer enhancement in a corona field by applying ionic wind. *Enhanc. Heat Transf.* **1997**, *4*, 71–86. [CrossRef]
13. Kawamoto, H.; Umezu, S. Electrostatic micro-ozone fan that utilizes ionic wind induced in pin-to-plate corona discharge system. *J. Electrostat.* **2008**, *66*, 445–454. [CrossRef]
14. Zhang, J.; Lai, F.C. Heat transfer enhancement using corona wind generator. *J. Electrostat.* **2018**, *92*, 6–13. [CrossRef]
15. Tsui, Y.-Y.; Huang, Y.-X.; Lan, C.-C.; Wang, C.-C. A study of heat transfer enhancement via corona discharge by using a plate corona electrode. *J. Electrostat.* **2017**, *87*, 1–10. [CrossRef]
16. Deylami, H.M.; Amanifard, N.; Dolati, F.; Kouhikamali, R.; Mostajiri, K. Numerical investigation of using various electrode arrangements for amplifying the EHD enhanced heat transfer in a smooth channel. *J. Electrostat.* **2013**, *71*, 656–665. [CrossRef]
17. Mahmoudi, A.R.; Pourfayaz, F.; Kasaeian, A. A simplified model for esitmating heat transfer coefficient in a chamber with electohydrodynamic effect (corona wind). *J. Electrostat.* **2018**, *93*, 125–136. [CrossRef]
18. Lakeh, R.B.; Molki, M. Enhancement of convective heat transfer by electrically-induced swirling effect in laminar and fully-developed internal flows. *J. Electrostat.* **2013**, *71*, 1086–1099. [CrossRef]
19. Lakeh, R.B.; Molki, M. Targeted heat transfer augmentation in circular tubes using a corona jet. *J. Electrostat.* **2012**, *70*, 31–42. [CrossRef]
20. Takahashi, K. *The Foundation of Aerosol Science*; Morikita Publishing Co., Ltd.: Tokyo, Japan, 2003.
21. Chen, J.; Davidson, J.H. Model of the negative dc corona plasma: Comparison to the positive dc corona plasma. *Plasma Chem. Plasma Process.* **2003**, *23*, 83–102. [CrossRef]
22. Farnoosh, N.; Adamiak, K.; Castle, G.S.P. 3-D numerical analysis of EHD turbulent flow and mono-disperse charge particle transport and collection in a wire-plate EHD. *J. Electrostat.* **2010**, *68*, 513–522. [CrossRef]
23. Kanamoto, H.; Yasuda, H.; Umezu, S. Flow distribution and pressure of air due to ionic wind in pin-to-plate corona discharge system. *J. Electrostat.* **2006**, *64*, 400–407. [CrossRef]
24. Tokyo Japan Society of Mechanical Engineers. *JSME Data Book: Heat Transfer*, 3rd ed.; Japan Society of Mechanical Engineers: Tokyo, Japan, 1976.
25. Cobine, J.D. *Gaseous Conductors: Theory and Engineering Applications*; Dover Publications Inc.: New York, NY, USA, 1957; p. 606.

Article

Collection Characteristic of Nanoparticles Emitted from a Diesel Engine with Residual Fuel Oil and Light Fuel Oil in an Electrostatic Precipitator

Akinori Zukeran *, Hidetoshi Sawano and Koji Yasumoto

Department of Electrical and Electronic Engineering, Kanagawa Institute of Technology,
Kanagawa 243-0292, Japan
* Correspondence: zukeran-akinori@ele.kanagawa-it.ac.jp

Received: 29 June 2019; Accepted: 24 August 2019; Published: 28 August 2019

Abstract: The purpose of this study was to investigate the collection characteristics of nanoparticles emitted from a diesel engine in an electrostatic precipitator (ESP). The experimental system consisted of a diesel engine (400 cc) and an ESP; residual fuel oil and light fuel oil were used for the engine. Although, the peak value of distribution decreased as the applied voltage increased due to the electrostatic precipitation effect, the particle concentration, at a size of approximately 20 nm, increased compared with that at 0 kV, in the exhaust gas, from the diesel engine with residual fuel oil. However, the efficiency increased by optimizing the applied voltage, and the total collection efficiency in the exhaust gas, using the residual fuel oil, was 91%. On the other hand, the particle concentration, for particle diameters smaller than 20 nm, did not increase in the exhaust gas from the engine with light fuel oil.

Keywords: diesel engine; ion-induced nucleation; collection efficiency; nanoparticle; electrostatic precipitator

---

1. Introduction

Although a renewable energy sources, such as biodiesel are effective in preventing global warming, it is necessary to remove particle emission. Nanoparticles are included in various diesel exhaust gases from diesel automobiles [1], power generation engines, marine diesel engines, and construction machines. These particles can penetrate into alveoli and are harmful to human health. Thus, electrostatic precipitators (ESPs) have been used or developed to negate these effects.

ESPs have been used to clean gas emissions from diesel automobiles in road tunnels [2]. The collection efficiency estimated by dust weight reached 90% at a gas speed of 9 m/s. Ehara et al. [3] investigated the collection of nanoparticles emitted from a diesel engine, with light fuel oil, in order to improve an ESP for road tunnels, and the efficiency for nanoparticles, with a size between 20 and 800 nm, was greater than 90% at a gas speed of 10 m/s.

ESPs have also been developed for power generators, farm engines, construction machines, and marine diesel engines. An ESP combined with an after-cyclone dust collector [4], a system with a mechanical filter located after an ESP [5], and an electrostatic cyclone diesel particulate filter (DPF) for marine engines [6], were proposed. However, the nanoparticle collection efficiency was not investigated. Kuroki et al. [7] reported the removal performance of nanoparticles (polystyrene latex) in a wet-type discharge plasma reactor, and showed a high collection efficiency greater than 99%. Mizuno et al. [8,9] investigated the effects of gas temperature and the addition of a DPF on collecting diesel nanoparticles in an ESP. Kim et al. [10–12] investigated the submicron particulate matter (PM) removal of an ESP, combined with a metallic filter for diesel engines and the nanoparticle collection of a novel two-stage type ESP, using KCl particles. Yamamoto and Ehara [13,14] suggested an electrohydrodynamic (EHD) assisted ESP, and a hole-type ESP, respectively. They measured the collection efficiency for nanoparticles emitted from diesel engines with light fuel oil. Authors have investigated nanoparticle collection efficiency for exhaust gas from a diesel engine with residual fuel oil [15]. As described so far, there are some studies on nanoparticle collection efficiency. However, the influence of the kind of fuel oil, applied voltage on nanoparticle collection efficiency, and efficiency for particles smaller than 20 nm, were not fully investigated.

In this study, we carried out an experiment using exhaust gases from a diesel engine, with residual and light fuel oils, to clarify the collection characteristics of nanoparticles in an ESP.

2. Experimental Setup

The schematic of the experimental system is shown in Figure 1. The system consists of a diesel engine (DA-3100SS-IV, 400 cc, 5.5 kW output, 0% load; Denyo Co., Ltd., Tokyo, Japan) and an ESP. Residual fuel oil (ENEOS LSA; sulfur content: 0.61%) and light fuel oil (ENEOS; sulfur content: 0.0009%) were used. The temperature of the exhaust gas was between 130 °C and 150 °C, and the gas temperature inside the ESP was between 70 and 80 °C. The wind velocity inside the ESP was 2.4 m/s.

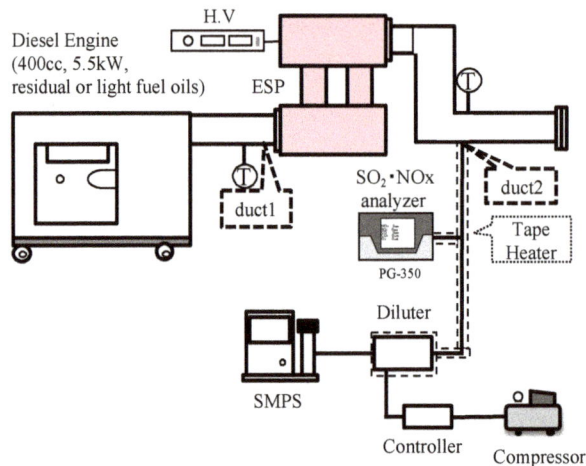

Figure 1. Schematic diagram of experimental system.

Figure 2 shows the structure of the ESP. It has a coaxial cylinder structure consisting of the grounded case (stainless; length: 80 mm; inner diameter: 58 mm) and high-voltage application wire electrode (tungsten; diameter: 0.26 mm). Negative or positive DC high voltage is applied to the wire electrode.

A portion of the gas in the duct (2.7 L/min) was drawn to measure the particle size distribution. The temperature of the sampling tube was controlled using the heater so that it would be equal to the temperature of the gas in the duct, in order to prevent cooling of the sampling tube, which could cause condensation inside the tube and change the rate of the components. The sampled gas was cooled to room temperature, after 10-fold dilution by the diluter (Palas KHG-2010 heatable dilution system), at the same temperature as the exhaust gas. It is considered that the vapor content in a gas is diluted to less than the saturated vapor content after the gas is cooled, whereby condensation hardly occurs. Although the diluter and dilution ratio were different from this equipment, Kim et al. [16,17] used a similar sampling and measurement system. The particle size distribution was measured using a scanning mobility particle sizer (SMPS) (Model 3936, TSI, Shoreview, MN, USA). The SMPS can measure the particle concentration for diameters between 6 nm and 200 nm. The collection efficiency η was calculated by Equation (1),

$$\eta = \left\{1 - \left(\frac{N}{N_0}\right)\right\} \times 100\% \tag{1}$$

where N is the particle number concentration (parts/m^3) after applying the voltage, and N_0 is the particle number concentration (parts/m^3) before applying the voltage.

H.V. wire electrode
(Tungsten wire, Φ 0.26 mm)

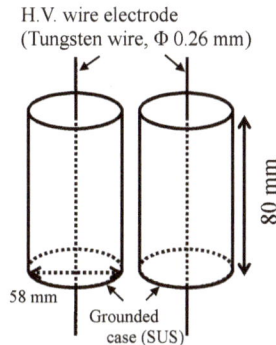

Figure 2. Structure of the electrostatic precipitator (ESP).

To measure the SO$_2$ and NOx concentrations using an analyzer (PG-350, Horiba, Kyoto, Japan), a portion of the exhaust gas was drawn from the duct and diluted by the diluter at the same temperature as the exhaust gas. The SO$_2$ concentration was measured by the infrared absorbing method, with a measurement limit of 2 ppm. The NOx concentration was measured by the chemiluminescence detector, with a measurement limit of 2.5 ppm.

3. Results and Discussion

3.1. Influence of Fuel Oil on Distribution

The relationship between the applied voltage and the discharge current in the exhaust gas, when residual and light fuel oil were used, is shown in Figure 3. The polarity of the voltage was negative. The corona onset voltage was approximately 5 kV, and the spark voltage was 15 kV or 16 kV at the residual fuel oil. Although the corona onset voltage and spark voltage at the light fuel oil were almost the same, the discharge current was slightly greater than that at the residual fuel oil. The cause of this requires future study.

The size distribution of various applied voltages in negative polarity when residual fuel oil is used is shown in Figure 4. The distribution at a voltage of 0 kV had a peak value of 7.0×10^{12} parts/m^3 at a size of 76 nm. The peak value decreased as the applied voltage increased, and that was 2.6×10^{11} parts/m^3 at 10 kV, due to the electrostatic precipitation effect. The collection efficiency was approximately 96%. However, the concentration at a size of 16 nm at 8 kV increased to 7.0×10^{11} parts/m^3, and at a size of 20 nm at 10 kV increased to 3.4×10^{12} parts/m^3.

Thus, the experiment was carried out using the light fuel oil to investigate the reason for this. The particle size distribution for various applied voltages when light fuel oil was used is shown in Figure 5. The distribution at a voltage of 0 kV had a peak of 2.3×10^{12} parts/m^3 at a size of 79 nm. The particle concentration at 79 nm decreased to 4.2×10^{11} parts/m^3, due to the electrostatic precipitation effect at a voltage of 10 kV. Furthermore, the particle concentration of approximately 20 nm did not increase, compared with Figure 4.

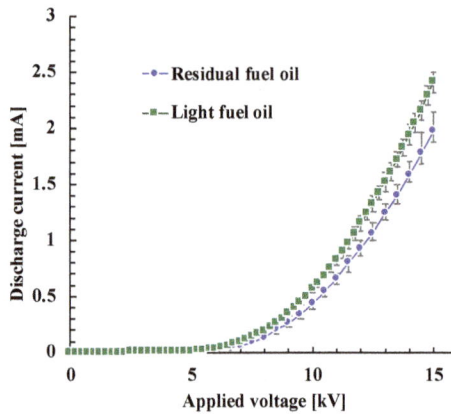

Figure 3. Relationship between applied voltage and discharge current when residual oil and light oil are used.

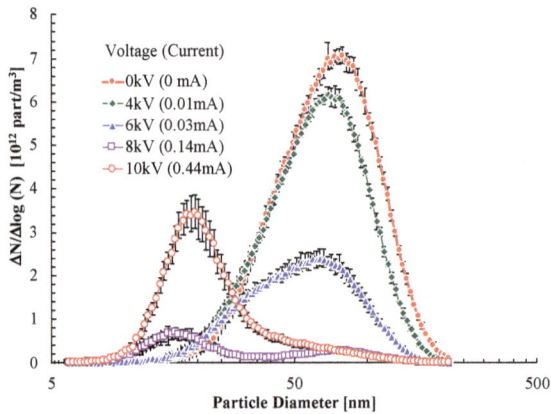

Figure 4. Size distribution of various applied voltages when residual oil is used.

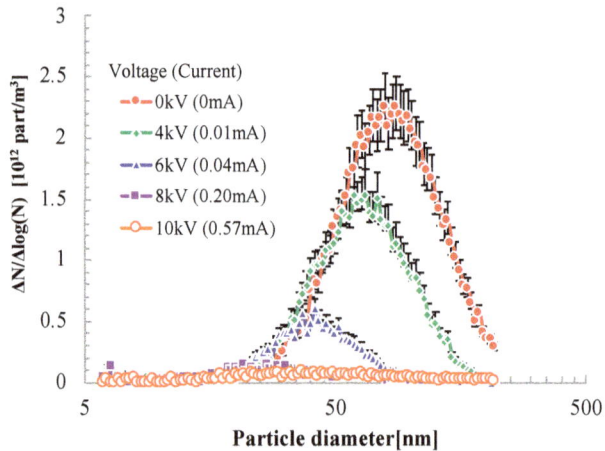

Figure 5. Size distribution for various applied voltages when light oil is used.

It is known whether nanoparticles are generated through a binary homogeneous nucleation and ion-induced nucleation in corona discharge [18,19]. Kim et al. [20,21] showed that particle concentration increased due to the ion-induced nucleation, caused by H_2O and the binary homogeneous nucleation from the effects of SO_2. H_2SO_4/H_2O particles were generated due to the interaction between SO_2 and OH radicals generated by the ionization of H_2O. This influence increased with increasing SO_2 and H_2O concentrations [19,22], as well as the addition of NO_2 [22]. On the other hand, Yohaness et al. [23] showed that the particle concentration, generated by corona discharge, decreased when NO_2 was added to $SO_2/H_2O/air$ mixed gas. Thus, the SO_2 and NOx concentrations were measured in this study. The analysis of the components in the exhaust gas revealed that the SO_2 concentration in the exhaust gas, using residual fuel oil, was 37 ppm, whereas it was less than 2 ppm in the case of light fuel oil. However, the measurement limit of the SO_2 analyzer was 2 ppm. The NOx concentration in the exhaust gas, using residual fuel oil, was 149 ppm, and the concentration using light fuel oil was 120 ppm. Therefore, the increased concentration for a particle size of approximately 20 nm in the exhaust gas, using residual fuel oil, may be due to the generation of H_2O particles and H_2SO_4/H_2O particles. H_2SO_4 molecules are generated from H_2O and OH^- in corona discharge. H_2SO_4 molecules easily convert to H_2SO_4/H_2O particles, due to binary homogeneous nucleation when H_2O particles, which are generated by ion-induced nucleation, are present in the gas.

3.2. Influence of Polarity on Distribution

It has been shown in Figure 4 that the particle concentration, at a size of approximately 20 nm, increased due to corona discharge. Thus, the influence of the polarity of the applied voltage in the exhaust gas, when residual fuel oil is used, was investigated.

The relationship between the applied voltage and the discharge current, when negative and positive voltages were individually applied, is shown in Figure 6. The corona onset voltage at the negative corona discharge was approximately 5 kV, and the spark voltage was between 15 kV and 16 kV. The onset voltage at the positive corona discharge was approximately 6 kV, and the spark was generated at 13 kV or 14 kV. Compared with the result of the negative corona discharge, the corona onset voltage was greater, and the spark voltage was smaller, at the positive corona discharge.

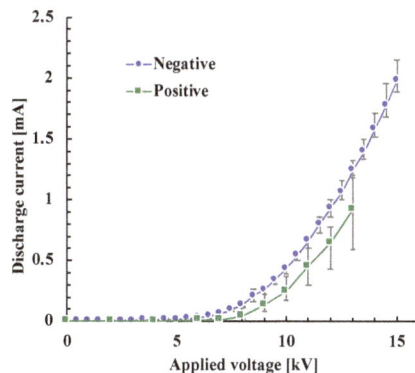

Figure 6. Relationship between applied voltage and discharge current (residual oil).

Although the peak value of the size distribution in negative polarity, at a voltage of 0 kV, decreased as the applied voltage increased, and the concentration at 20 nm size increased, as shown in Figure 4. The size distribution for various applied voltages in positive polarity is shown in Figure 7. The overall tendency was similar to the result in negative polarity, as shown in Figure 4. However, nanoparticle generation occurred from 8 kV at negative polarity, but it did not occur at 8 kV at the positive polarity.

This is considered to be caused by the corona discharge current at the positive polarity, less than that at the negative polarity, as shown in Figure 6.

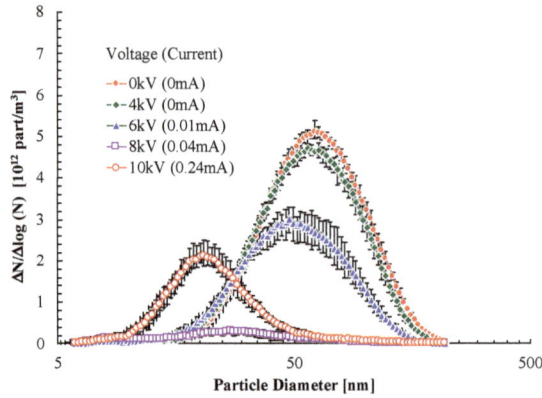

Figure 7. Size distribution for various applied voltages in positive polarity (residual oil).

A comparison between the results of negative and positive polarities, at a voltage of 10 kV, is shown in Figure 8. Figure 8a shows a comparison of the size distributions. In this Figure, the size distributions of both polarities, at 0 kV were not the same, so it is difficult to investigate the influence of polarity on the increased particle concentration of approximately 20 nm. Therefore, the enhancement rate α is defined as Equation (2):

$$\alpha = N/N_0 \qquad (2)$$

where N is the particle number concentration (parts/m^3) after applying the voltage, and N_0 is the particle number concentration (parts/m^3) before applying the voltage.

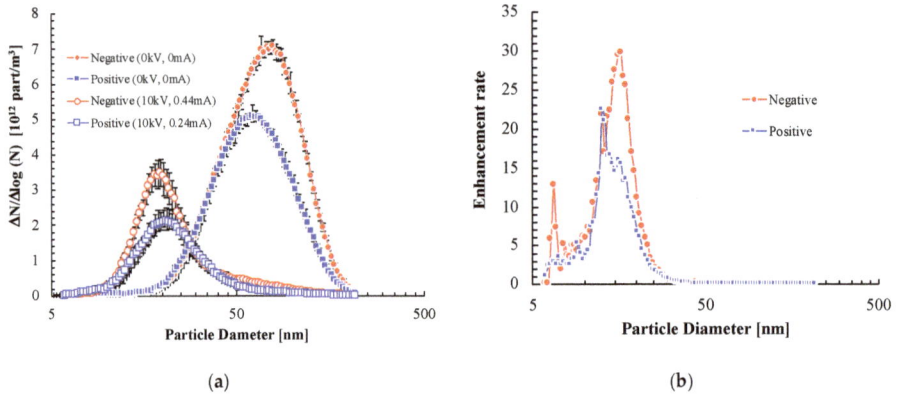

(a) (b)

Figure 8. Comparison between results of negative and positive polarities: (**a**) size distribution; (**b**) enhancement rate.

Figure 8b shows a comparison of the enhancement rates between negative and positive polarities. The enhancement rate was calculated from the average concentrations, as shown in Figure 8a, thus the error bar is not indicated in Figure 8b. The maximum enhancement rate at negative polarity was 30 at a size of 16 nm, and that at positive polarity was 22 at a size of 13 nm. This result indicates that the increased nanoparticles, due to the corona discharge in negative polarity, is greater than that in positive polarity. Nagato et al. [19] reported that the increase of particles in positive polarity was greater than that in negative polarity, which was different from our result. Although, their experiment was carried out with the same discharge current, the experiment in this study was at the same voltage. This is most likely option, because the discharge current in the negative corona is greater than that in the positive corona at the same voltage, as shown in Figure 6. The reason why sharp peaks occurred at different size ranges needs to be investigated in the future.

3.3. Collection Efficiency

An increase of fine particles is undesirable for an ESP. Thus, the effects of the applied voltage were investigated. The polarity of the voltage was negative, where the residual fuel oil was used.

The collection efficiency, as a function of particle diameter for various applied voltages, is shown in Figure 9. The collection efficiency, at a voltage of 10 kV, decreased with decreased particle diameter, and had a minimum value of approximately −2900% at a particle size of 16 nm. A negative collection efficiency means that the particle number concentration after the application of voltage is greater than before. Thus, the concentration at 16 nm increased by 30 times after the application of 10 kV. However, the efficiency at a size of approximately 16 nm increased as the applied voltage decreased. Kim et al. [16,17] reported that a nanoparticle collection efficiency of greater than 90% was achieved in an electrostatic filtration system, combined with a metallic flow-through filter for diesel exhaust gas, which was different from the result in this study. This is the most likely option, as their electrostatic filtration system, combined with a metallic flow-through filter, may have been suitable for preventing an increase of nanoparticles.

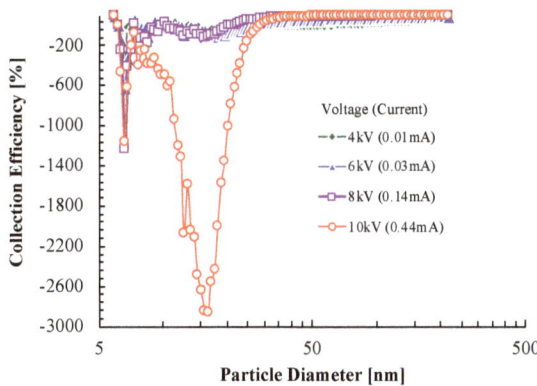

Figure 9. Collection efficiency as a function of particle diameter for various applied voltages in negative polarity (residual oil).

The total collection efficiency value, as a function of applied voltage, is shown in Figure 10. The efficiency increased as the applied voltage increased, and reached 91% at 8 kV, while the efficiency at 10 kV was 65%. This result shows that an optimum voltage exists for collection efficiency. This is probably due to ion-induced and binary homogeneous nucleation, as already described.

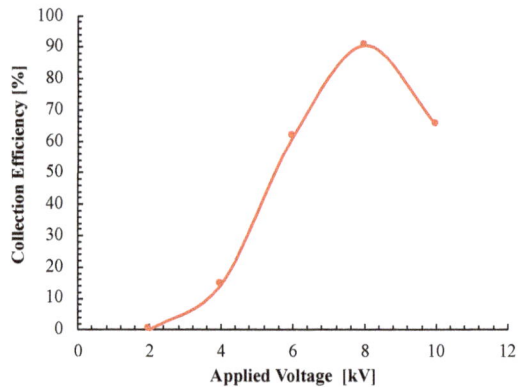

Figure 10. Collection efficiency as a function of applied voltage in negative polarity (residual oil).

4. Conclusions

The experiments were conducted to investigate the collection characteristics of nanoparticles, emitted from a diesel engine, with a residual fuel oil and light fuel oil in the ESP. The results are follows:

(1) The peak concentration, at 76 nm of size distribution in the exhaust gas from residual fuel oil, decreased with an in increase in the applied voltage. However, the nanoparticle concentration at a size of 20 nm increased. On the other hand, the nanoparticle concentration did not increase in the exhaust gas when light fuel oil was used. These results indicate that the increased nanoparticle concentration in the exhaust gas, using residual fuel oil, may be due to ion-induced and binary homogeneous nucleation.

(2) The amount of nanoparticle increase due to corona discharge in negative polarity was greater than that in positive polarity at the same voltage.

(3) An optimum voltage, used to suppress nanoparticle concentration, exists for the collection efficiency.

Author Contributions: Conceptualization, A.Z.; Formal analysis, A.Z. and H.S.; Funding acquisition, A.Z.; Investigation, A.Z., H.S. and K.Y.; Methodology, A.Z., H.S. and K.Y.; Project administration, A.Z.; Writing – original draft, A.Z.; Writing – review & editing, A.Z., H.S. and K.Y.

Funding: This work was supported by a Grant-in-Aid for Scientific Research (B), Nos. 15H04216 and 18H01647, from the Japan Society for the Promotion of Science.

Conflicts of Interest: The authors declare no conflict of interest.

References

1. Zukeran, A.; Ikeda, Y.; Ehara, Y.; Matsuyama, M.; Ito, T.; Takahashi, T.; Kawakami, H.; Takamatsu, T. Two-Stage Type Electrostatic Precipitator Re-entrainment Phenomena under Diesel Flue Gases. *IEEE Trans. Ind. Appl.* **1999**, *35*, 346–351. [CrossRef]

2. Katatani, A.; Dix, A. *Ventilation and Exhaust Purification of Motor Vehicle Tunnels in Japan*; BHR Group 2011 ISAVT14: Bedford, UK, 2011; pp. 577–588.

3. Ehara, Y.; Nakano, R.; Yamamoto, T.; Zukeran, A.; Inui, T.; Kawakami, H. Performance of High Velocity Electrostatic Precipitaor for Road Tunnel. *Int. J. Plasma Environ. Sci. Technol.* **2011**, *5*, 157–160.

4. Isahaya, F. Development on electrostatic pre-coagulator combined with after-cyclone dust collector. *Hitachi Hyoron* **1967**, *49*, 77–80.

5. Masuda, S.; Moon, J.D.; Aoi, K. AUT—AINER Precipitator System—An Effective Control Means for Diesel Engine Particulates. Actas 5. *Congreso Int aire Pure 1980* **1982**, *2*, 1149–1153.

6. Sasaki, H.; Tsuamoto, T.; Furugen, M.; Makino, T. Reduction of PM emission from 4-stroke marine diesel engine by electrostatic cyclone DPF. *J. JIME* **2010**, *45*, 139–145.

7.   Kuroki, T.; Nishii, S.; Okubo, M. Fundamental Study on the Simultaneous Removal of Nanoparticles and Harmful Gas Components Using a Wet-Type Discharge Plasma Reactor. *Earozoru Kenkyu* **2015**, *30*, 108–113. (In Japanese)

8.   Takasaki, M.; Kubota, T.; Hayashi, M.; Kurita, H.; Takashima, K.; Mizuno, A. Electrostatic Precipitation of diesel PM at reduced gas temperature. In Proceedings of the 2015 IEEE Industry Applications Society Annual Meeting, Addison, TX, USA, 18–22 October 2015; pp. 1–4.

9.   Hayashi, H.; Takasaki, Y.; Kawahara, K.; Takenaka, T.; Takashima, K.; Mizuno, A.; Chang, M.B. Electrostatic charging and precipitation of diesel soot. In Proceedings of the 2009 IEEE Industry Applications Society Annual Meeting, Houston, TX, USA, 4–8 Octobeer 2009; pp. 1–8.

10. Kim, H.J.; Han, B.; Woo, C.G.; Kim, Y.J. Submicron PM Removal of an ESP Combined with a Metallic Foam Filter for Large Volumetric Diesel Engines. *IEEE Trans. Ind. Appl.* **2015**, *51*, 4173–4179. [CrossRef]

11. Kim, H.J.; Han, B.; Woo, C.G.; Kim, Y.J. Performance of Ultrafine Particle Collection of a Two-Stage ESP Using a Novel Mixing Type Carbon Brush Charger and Parallel Collection Plates. *IEEE Trans. Ind. Appl.* **2017**, *53*, 466–473. [CrossRef]

12. Kim, H.J.; Han, B.; Woo, C.G.; Kim, Y.J. Ultrafine Particle Collection Performance of a Two-Stage ESP with a Novel Geometries and Electrical Conditions. *IEEE Trans. Ind. Appl.* **2017**, *53*, 5859–5866. [CrossRef]

13. Kawakami, H.; Sakurai, T.; Ehara, Y.; Yamamoto, T.; Zukeran, A. Performance characteristics between horizontally and vertically oriented electrodes EHD ESP for collection of low-resistive diesel particulates. *J. Electrost.* **2013**, *71*, 1117–1123. [CrossRef]

14. Ehara, Y.; Ohashi, M.; Zukeran, A.; Kawakami, K.; Inui, T.; Aoki, Y. Development of Hole-Type Electrostatic Precipitator. *Int. J. Plasma Environ. Sci. Technol.* **1017**, *11*, 9–12.

15. Kawakami, H.; Zukeran, A.; Yasumoto, K.; Inui, T.; Ehara, Y.; Yamamoto, T. Diesel PM Collection for Marine Emissions Using Double Cylinder Type Electrostatic Precipitator. *Int. J. Plasma Environ. Sci. Technol.* **2011**, *5*, 174–178.

16. Kim, H.J.; Han, B.; Hong, W.S.; Shin, W.H.; Cho, G.B.; Lee, Y.K.; Kim, Y.J. Development of Electrostatic Diesel Particulate Matter Filtration Systems Combined with a Metallic Flow-Through Filter and Electrostatic Method. *Int. J. Automot. Technol.* **2010**, *11*, 447–453. [CrossRef]

17. Kim, H.J.; Han, B.; Cho, G.B.; Kim, Y.J.; Yoo, J.S.; Oda, T. Collection Performance of an Electrostatic Filtration System Combined with a Metallic Flow-Through Filter for Ultrafine Diesel Particulate Matters. *Int. J. Automot. Technol.* **2013**, *14*, 489–497. [CrossRef]

18. Adachi, M.; Kusumi, M.; Tsukui, S. Generation of Nanodroplets and Nanoparticles by Ion-Induced Nucleation. *J. Soc. Powder Technol. Jpn.* **2004**, *41*, 424–430. (In Japanese) [CrossRef]

19. Nagato, K.; Yoshizumi, H.; Nonaka, Y.; Fukagawa, K. Effect of discharge polarity on the fine particle formation from SO_2 by DC corona discharge. *Earozoru Kenkyu* **2008**, *23*, 101–107. (In Japanese)

20. Kim, T.O.; Adachi, M.; Okuyama, K.; Seinfeld, J.H. Experimental Measurement of Competitive Ion-Induced and Binary Homogeneous Nucleation in $SO_2/H_2O/N_2$ Mixtures. *Aerosol Sci. Technol.* **1997**, *26*, 527–543. [CrossRef]

21. Kim, T.O.; Adachi, M.; Okuyama, K.; Seinfeld, J.H. Nanometer Sized Particle Formation from $NH_3/SO_2/H_2O$/Air Mixtures by Ionizing Irradiation. *Aerosol Sci. Technol.* **1998**, *29*, 111–125. [CrossRef]

22. Adachi, M.; Kim, C.S.; Kim, T.O.; Okuyama, K. Effects of NO_2 Gas on Gas-to-Particle Conversion of SO_2 by a-Ray Radiolysis. *Kagaku Kougaku Ronbunshu* **1999**, *25*, 868–872. (In Japanese) [CrossRef]

23. Yohannes, P.; Xiaoping, B.; Stelson, A.W. Competition of NO and SO_2 for OH Generated within Electrical Aerosol Analyzers. *Aerosol Sci. Technol.* **1995**, *22*, 190–193. [CrossRef]

energies

MDPI

Article

High Reduction Efficiencies of Adsorbed NOx in Pilot-Scale Aftertreatment Using Nonthermal Plasma in Marine Diesel-Engine Exhaust Gas

Takuya Kuwahara [1],*, Keiichiro Yoshida [2], Tomoyuki Kuroki [3], Kenichi Hanamoto [4], Kazutoshi Sato [4] and Masaaki Okubo [3]

[1]   Department of Mechanical Engineering, Nippon Institute of Technology, 4-1 Gakuendai, Miyashiro-machi, Minamisaitama, Saitama 345-8501, Japan
[2]   Department of Electrical and Electronic Systems Engineering, Osaka Institute of Technology, 5-16-1 Omiya, Asahi-ku, Osaka 535-8585, Japan; keiichiro.yoshida@oit.ac.jp
[3]   Department of Mechanical Engineering, Osaka Prefecture University, 1-1 Gakuen-cho, Naka-ku, Sakai 599-8531, Japan; kuroki@me.osakafu-u.ac.jp (T.K.); mokubo@me.osakafu-u.ac.jp (M.O.)
[4]   Daihatsu Diesel MFG. Co., Ltd., 45 Amura-cho, Moriyama city, Shiga 524-0035, Japan; kenichi.hanamoto@dhtd.co.jp (K.H.); kazutoshi.sato@dhtd.co.jp (K.S.)
*   Correspondence: takuya.k@nit.ac.jp; Tel.: +81-480-33-7719

Received: 4 September 2019; Accepted: 4 October 2019; Published: 8 October 2019

Abstract: An efficient NO_x reduction aftertreatment technology for a marine diesel engine that combines nonthermal plasma (NTP) and NO_x adsorption/desorption is investigated. The aftertreatment technology can also treat particulate matter using a diesel particulate filter and regenerate it via NTP-induced ozone. In this study, the NO_x reduction energy efficiency is investigated. The investigated marine diesel engine generates 1 MW of output power at 100% engine load. NO_x reduction is performed by repeating adsorption/desorption processes with NO_x adsorbents and NO_x reduction using NTP. Considering practical use, experiments are performed for a larger number of cycles compared with our previous study; the amount of adsorbent used is 80 kg. The relationship between the mass of desorbed NO_x and the energy efficiency of NO_x reduction via NTP is established. This aftertreatment has a high reduction efficiency of 71% via NTP and a high energy efficiency of 115 g(NO_2)/kWh for a discharge power of 12.0 kW.

Keywords: aftertreatment; energy efficiency; marine diesel engine; nonthermal plasma; NO_x

---

1. Introduction

The advantages of diesel engines are low CO_2 emissions and high fuel efficiency with respect to the output power. In general, ships use diesel engines as propulsion and auxiliary engines because various types of fuels can be employed. However, their emissions contain harmful particulate matter (PM) and NO_x (NO + NO_2). Therefore, exhaust purification requires aftertreatment technologies [1]. These technologies have been extensively studied and developed in recent years, e.g., in our previous studies [2,3].

Reduction of NO_x from diesel emissions is difficult. However, emissions requirements have become increasingly stringent in recent years. The improvement of fuel injection in these engines has been investigated [4]. To reduce NO_x emissions in O_2-rich environments, selective catalytic reduction (SCR) using urea solution or ammonia has been utilized [5–11]. Some methods combine SCR with nonthermal plasma (NTP) [12–16]. However, SCR requires a high temperature of 300 °C for catalyst activation, and there are issues regarding the production of nanoparticles, leakage of harmful ammonia, use of harmful heavy-metal SCR catalysts, and storage of the urea solution. Additionally, NO_x reduction technologies using NTP without a catalyst have been investigated [17,18].

Marine diesel engines are the next target of regulations. Normally, NO_x concentrations in the emissions of marine diesel engines are relatively high (500–1000 ppm), and the emissions contain SO_x (typically 100 ppm) in addition to PM [2].

The regulations that govern marine diesel engines are specified by the International Maritime Organization (IMO) as explained for a rotation speed of 900 rpm in [2]. In these regulations, the emission is restricted by the mass of NO_x emitted per unit of engine output energy for a given engine rotation speed, and is divided into Tiers I, II, and III according to the enforcement period. For a marine diesel engine with a rotation speed of 900 rpm, which is the target engine type in this study, NO_x emissions should be <11.5 g/kWh for Tier I in 2000, 9.20 g/kWh for Tier II in 2011, and 2.31 g/kWh in a NO_x-regulated emission area for Tier III in 2016. Therefore, a NO_x reduction of 6.89 g/kWh is required from Tier II to the current Tier III applied in 2016. This corresponds to a 75% reduction. Given the importance of NO_x emissions, more stringent regulations may be imposed on marine diesel engines in the future, which could require a similar amount of NO_x reduction.

Considering the circumstances of marine diesel emission and the emission regulations, NO_x treatment, as well as the removal of PM, must be urgently addressed [19–22]. In 2011, the main manufacturers of marine diesel engines were able to successfully satisfy the Tier II NO_x emission standards via combustion improvements. However, to satisfy the more stringent Tier III, an effective aftertreatment technology is indispensable. Urea-SCR technology is presently the most promising approach. However, it requires the storage of large amounts of urea solution inside the vessels. NO_x reduction via wet scrubbing using a chemical solution has also been investigated [23]. However, it requires a tank of chemical solution in the ship, which has limited space.

In this study, an aftertreatment technology for NO_x reduction and PM treatment using NTP for a marine diesel engine is developed on the basis of our previous studies in a laboratory-scale experiment [24,25]. Considering practical use, experiments are performed for a larger number of cycles compared with our previous study to obtain more data. The objective is to obtain a more accurate relationship between the mass of desorbed NO_x and the mass of NO_x reduced by NTP with the data. The amount of adsorbent is 80 kg. Compared with SCR, this technology offers the advantages of eliminating the requirement of urea solution or harmful heavy-metal catalysts and operation at a temperature of <150 °C. In our previous investigation [3], PM and NO_x reductions were studied in the aftertreatment for a marine diesel engine with an output power of 610 kW. In another one of our previous studies [2], NO_x reduction with NO_x adsorbents unused in the aftertreatment was investigated for a marine diesel engine with an output power of 1071 kW. In the present study, according to the results of our previous investigations regarding NO_x reduction [2], experiments involving aftertreatment for a marine diesel engine with an output power of 1071 kW are repeated for a longer period (up to 19 cycles). In the aftertreatment, estimation of the NO_x reduction efficiency is significant for the design and operation of the system. Compared with the previous studies, considerably more data are obtained to provide a highly accurate estimation of the efficiency of NO_x reduction via NTP.

2. Operating Principle of NO_x Reduction in Aftertreatment

In the previously reported technology [2,24,25], given that NO_x cannot be efficiently and directly reduced by NTP under O_2-rich conditions, it is first adsorbed by adsorbents under O_2-rich conditions. After the adsorption, NO_x is desorbed by heating the adsorbents under O_2-lean (preferably $O_2 < 2\%$) and fuel-rich (CO and hydrocarbon-rich) gaseous conditions. Switching the different processes is achieved by changing the exhaust path flows and by using the waste-heat recovery of the engine. The high-concentration NO_x desorbed from the adsorbents is effectively reduced to N_2 and N radicals by NTP. O_2-lean or N_2 gaseous conditions can be achieved using an O_2 penetration membrane or by controlling the engine operating mode (fuel-injection mode). The following chemical reaction for NO_x reduction is performed under O_2-lean conditions with NTP:

$$2NO_x + 2N \rightarrow 2N_2 + xO_2 \ (>\text{room temperature}). \tag{1}$$

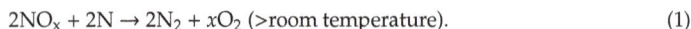

The NTP reactors should only be turned on when high concentrations of NO_x are desorbed and only during the short desorption period, reducing the required plasma energy. Laboratory-scale experiments based on this procedure in which NO_x was treated with a high energy efficiency were reported in our previous papers [26–29], and a related patent was filed [30].

Figure 1 shows process diagrams of the PM and NO_x simultaneous reduction system for a marine diesel engine. Because the objective of the present study is to investigate the NO_x reduction efficiencies in a larger number of cycles compared with our previous study [2], the experimental setup and processes are the same. The system mainly consists of the marine diesel engine, diesel particulate filters (DPFs), an adsorption chamber, and NTP reactors. PM reduction is first performed with a pair of DPFs, and the DPFs are regenerated via ozone injection. Next, NO_x reduction is achieved through two flow processes: an adsorption process followed by a desorption process combined with the NTP reaction. The sequential application of these two processes in the same adsorption chamber realizes continuous NO_x reduction. In the adsorption process, the flow rate of the exhaust gas determines the flow velocity, which is measured by pitot tubes. After the exhaust gas is cooled using a water-cooling-type cooler, it passes through the adsorption chamber, where NO_x is adsorbed by adsorbents. The NO_x concentrations are measured at the inlet and outlet of the chamber. In the desorption process, the exchanged waste heat is added to the adsorbent pellets via a heat exchanger to induce the thermal desorption of NO_x. Simultaneously, N_2 gas flows over the packed adsorbent pellets. Then, high concentrations of NO_x are desorbed from the chamber. The desorbed NO_x is reduced in the NTP reactor. Thus, the simultaneous reduction of PM and NO_x is achieved [2]. In the next section, the experimental apparatus and results are presented.

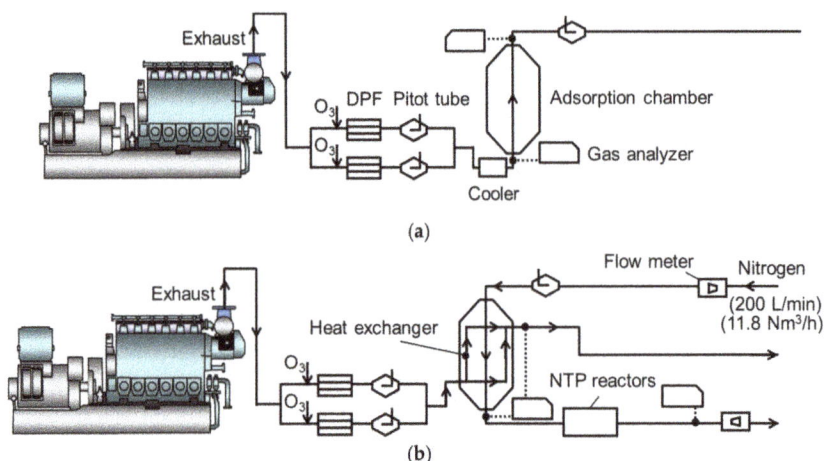

Figure 1. Process diagram of the NO_x reduction system for a marine diesel engine: (**a**) adsorption process; (**b**) desorption process.

3. Experimental Apparatus

A photograph of the diesel engine (6DK-20e, Daihatsu Diesel MFG Co. Ltd., Japan) is shown in Figure 2. Experiments are performed using an electrical sub-power generation marine engine bench in the laboratory. Table 1 shows the specifications of the engine. The specifications are four strokes, six cylinders with a cylinder bore of 200 mm and a stroke of 300 mm, and a constant rotation rate of 900 rpm. Table 2 shows the operating conditions. The maximum (100%) output power is 1071 kW. The fuel was marine diesel oil (A-heavy oil (the same grade as marine diesel oil), sulfur content = 0.075 mass%, nitrogen content = 0.01 mass%, heating value = 45.4 MJ/kg). The exhaust flow rate was 3920 Nm^3/h for 50% load or output power, 5526 Nm^3/h for 75%, and 6815 Nm^3/h for 100% (N denotes the standard state of 0 °C, 0.1 MPa) [2].

Figure 2. Photograph of the targeted marine diesel engine (6DK-20e, maximum power = 1071 kW).

Table 1. Specifications of the marine diesel engine (6DK-20e).

Specification	Value
Number of Cylinders	6
Bore, mm	200
Stroke of cylinder, mm	300
Rotating speed, rpm	900
Weight of engine with a dynamo, tons	16

Table 2. Operating conditions of the marine diesel engine (6DK-20e).

Specification	Value	Value	Value
Load, %	100	75	50
Power, kW	1071	803	536
Exhaust gas flow, Nm^3/h	6815	5526	3920
Air flow rate from supercharger, $Nm^3/kWmh$	6.36	6.88	7.32
Exhaust gas components			
NO_x, ppm	780	710	660
CO, ppm	105	63	33
CO_2, %	5.4	4.9	4.8
O_2, %	13.6	14.2	14.4
HCs, ppm	120	120	120

Figure 3 shows a schematic of the experimental setup for the aftertreatment for the marine diesel engine. Figure 4 shows the experimental setup for exhaust-gas aftertreatment in the marine diesel engine system. Approximately 16% of the bypassed exhaust gas passes into a 150A pipe (in Japanese Industrial Standards; inner diameter = 155.2 mm) and through a set of ceramic DPFs (material: SiC, TYK

Corporation, Japan). Here, most of the PM is removed. Subsequently, the flow velocity is measured by a set of pitot tubes (L type, FV-21A, OKANO WORKS, Ltd. Japan). The accumulated PM in the DPF is treated using NTP-induced ozone (O_3) injection technology, as we previously reported [2,31]. After the PM removal, the NO_x in the exhaust gas is treated via adsorption and desorption processes with NTP in the same way as the previous experiment [2] as well as the measurements. The concentration of untreated NO_x from the engine is 400-760 ppm, and the ratio of NO_2/NO_x is approximately 15%. N_2 gas at a low flow rate in the desorption process is 11.8 Nm^3/h (200 L/min at 5 °C). A significantly higher concentration (typically 3660-25,000 ppm) of NO_x compared with that of the previous study (typically 4000 ppm) [2] flows out of the chamber and enters the NTP reactors, with a total energy consumption of 12.0 kW. NO_x is reduced into N_2 and O_2 according to reaction (1). As the upper limit of the analyzer is 2500 ppm, high-concentration desorbed NO_x gas in excess of 2500 ppm is diluted with atmospheric air. The actual NO_x concentration is estimated by comparing the O_2 concentration of the diluted exhaust gas with that of the raw exhaust.

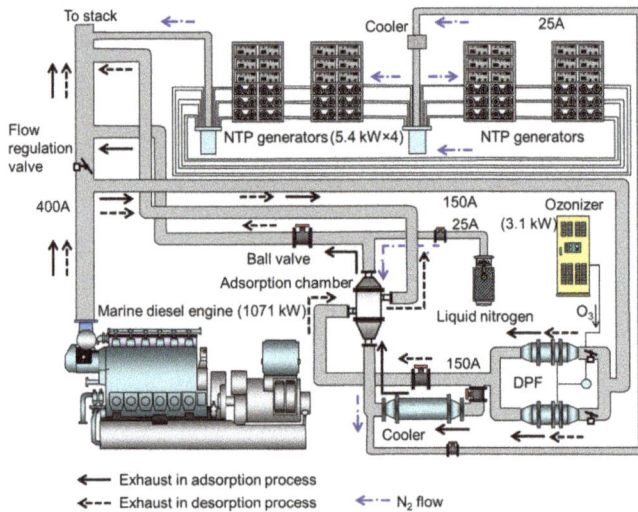

Figure 3. Schematic of the experimental setup for exhaust-gas aftertreatment in the marine diesel engine.

Figure 4. Photograph of the experimental setup for exhaust-gas aftertreatment in the marine diesel engine system.

Figure 5 shows the adsorption chamber equipped with a waste-heat exchanger that is specially designed and manufactured by Sumitomo Precision Products Co. Ltd. (type: XS6083). The directions of gas flow and the dimensions are shown. The same chamber that was used in a previous experiment is employed [2]. The adsorption chamber is designed optimally based on our results of the laboratory-scale experiment [25] and the previous pilot-scale ones [2,3]. The amount of packed adsorbent pellets is 80 kg, which represents 101 L by volume. Compared with the previous study [3], the dimensions of the chamber are different, and the cross-sectional area is 3.9 times larger. However, the vertical length is 0.7 times shorter. The volume of the adsorption chamber including relevant externals is approximately 0.5 m³, which is smaller than the typical volume of 6.0 m³ of a urea-solution tank. The mass of the adsorbent chamber without adsorbents is almost equal to that of empty urea-solution tank. Figure 5a shows a cross section of the chamber with two types of flow paths—flow path I (the number is 47, and each gap is 3.2 mm) and flow path II (the number is 48, and each gap is 8.9 mm)—alternately stacked inside. Figure 5b shows a side view of flow path I, in which the hot exhaust gas flows. Flow path I is empty, and flow path II is packed with adsorbent pellets, as shown in Figure 5c. In the adsorption process, while exhaust gas flows from the bottom inlet to the top outlet of flow path II, NO_x is adsorbed onto the pellets. In the desorption process, heated exhaust gas travels along flow path I to heat the adsorbent pellets. Simultaneously, N_2 gas from a liquid N_2 cylinder flows from the top inlet to the bottom outlet of flow path II at a low flow rate to achieve O_2-lean condition, as shown in Figure 5b,c. Switching between these two processes is performed by opening and shutting the ball valves. The adsorbent used in this study is a MnO_x–CuO oxidative compound (N-140, 1.2–2.4 mm-sized granular pellets, Süd-Chemie Catalysts Japan, Inc.). The measurement points for the adsorbent temperature are shown in Figure 5c. The temperatures measured at these points are averaged for evaluating the efficiencies.

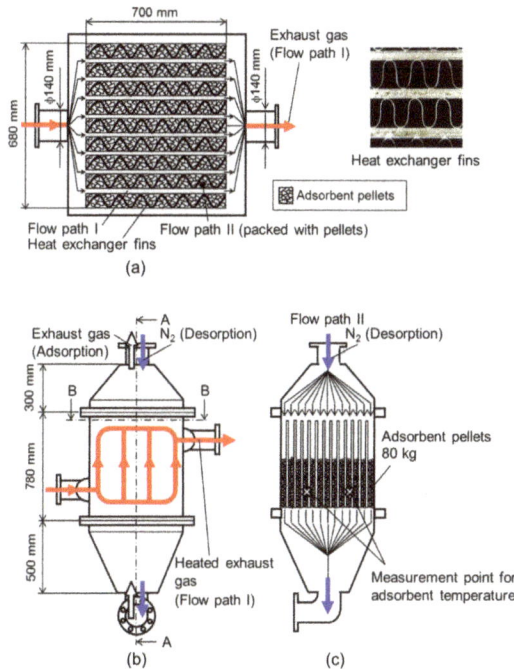

Figure 5. Schematic of the adsorption chamber containing 80 kg of adsorbents with gas flows in the desorption process. (**a**) Cross section B-B; (**b**) Side view; (**c**) Cross section A-A;

Table 3 presents the design specifications of the adsorption chamber. A counter-flow-type heat exchanger is used in the adsorption chamber. The design specifications are the same as those used in a previous study [2]. Thus, the total heat-exchange quantity is 61.2 kW. The pressure drop and space velocity are also presented in the table. When the amount of packed adsorbent pellets is 40 kg, the space velocity is higher than that in our previous study [3], with a ratio of 1.96 (i.e., 16,000/8150). Figure 6 shows a photograph and schematic of the NTP reactor used for reducing NO_x. The reactor consists of a surface-discharge element (ET-OC70G-C, Masuda Research Inc., Japan), air-cooling fins, and a flange to fix the discharge element to the frame. The structure of surface discharge is also presented in the figure. As shown, NO_x in the N_2 gas flows on the surface-discharge element. NO_x is reduced to the clean gases of N_2 and O_2 with the surface discharge plasma. The surface-discharge element is cooled with an air-cooling fan. The specifications of one unit of the NTP generator in Figure 3, which includes the power supplies and the NTP reactors, are as follows. Two of these reactors are powered by a single-pulse high-voltage power supply (HCII-70/2, Masuda Research Inc.). The maximum peak-to-peak voltage is 10 kV, with a frequency of 10 kHz. The maximum input power is 450 × 2 = 900 W. A unit of the NTP generator (HCII-OC70×12) consists of 12 NTP reactors and six power supplies. The total input power of a unit is 900 W × 6 = 5.4 kW, and the discharge power is 5.0 kW.

Table 3. Specifications of the adsorption chamber.

Specification	Value
Operation Gauge Pressure	0.1 MPa
Heatproof temperature	300 °C
Material of heat-exchanger fin	Stainless steel
Area and type of heat-exchanger fin Flow path I Flow path II	55 m², serrated fin 72 m², plain fin
Mass of adsorbent chamber without adsorbent pellets	920 kg
Amount and volume of packed adsorbent pellets	80 kg, 101 L
Pressure drop in adsorption process with adsorbent pellets (80 kg)	2 kPa (Load 75%, 810 Nm³/h)
Space velocity (exhaust gas: 800 Nm³/h, 175 °C; adsorbents: 80 kg)	16,000/h

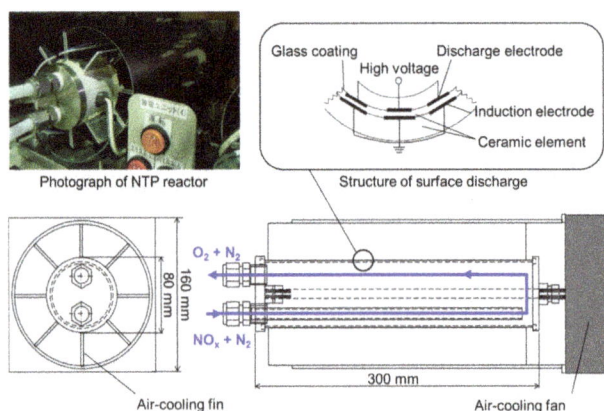

Figure 6. Schematic of the nonthermal plasma (NTP) reactor, showing the structure of the discharge section, along with a photograph of the NTP reactor.

4. Results and Discussion

Experiments are performed for 19 operation cycles. The NO_x reduction performance in the aftertreatment is evaluated. The engine operation was stopped once during each process.

Figure 7 shows the time-dependent NO_x emissions before and after the gas passes through the aftertreatment for cycles 16–19. Cycles 16–19 represent the transition of the adsorbent from the unsteady state to the steady state. The engine load is set as 75% for all adsorption processes and 50% for all desorption processes, considering the exhaust-gas temperatures for the adsorption and desorption of NO_x. This setting of the engine load is chosen to investigate the performance in a severe condition because it gives the severe condition for the aftertreatment, that is, high concentration of NO_x in adsorption and lower temperature in desorption processes. The amount of adsorbent pellets in the adsorption chamber is 80 kg. The mass flow rate for NO_x, which is shown on the vertical axis in Figure 7, is evaluated according to the molecular mass of NO_2, with the unit of g(NO_2)/h. Untreated NO_x in the adsorption process is represented by white circles with lines. Treated NO_x is represented by black circles with lines. NTP is applied only in the desorption processes, and the input power to the NTP generator is 12.0 kW. The mass flow rate of NO_x in the untreated exhaust gas is 1330–1500 g(NO_2)/h in the steady state of engine operation. The engine operation is stopped at $t = 3900$ min in the adsorption process of cycle 17 and at $t = 4192$ min in the adsorption process of cycle 18. Each stoppage lasts for approximately half a day. It is noted that the difference in the duration of the desorption is just due to the engine operation timing. In the adsorption processes, the mass flow rate of NO_x decreases to 970–1280 g(NO_2)/h. In the desorption processes, the maximum concentrations of desorbed NO_x are 8180, 11,380, 3660, and 17,830 ppm in cycles 16–19, respectively. On average, 49% of the desorbed NO_x is reduced by the application of NTP. For example, considering cycle 19 in the graph, similar to the previous report [2], the hatched area represents the total mass of adsorbed NO_x, and the area in the desorption process represents the total mass of NO_x reduced by the NTP. The desorption of NO_x is enhanced in cycle 19.

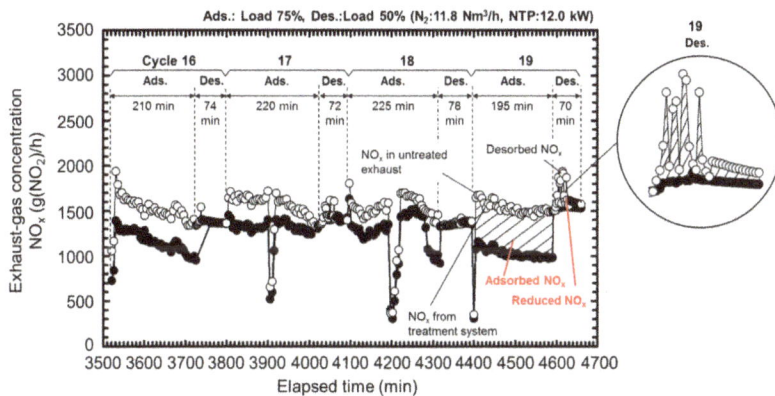

Figure 7. Time-dependent NO_x emissions before and after the gas passes through the aftertreatment for operation cycles 16–19.

Figure 8 shows the time-dependent temperature of the adsorbent pellets in cycles 16–19. At the beginning of each adsorption process, the adsorbent temperature is high because of residual heat from the previous desorption process. However, the temperature rapidly decreases to 50 °C under cooling. The exhaust gas is exceptionally uncooled at the beginning and is cooled at $t = 3606$ min in the adsorption process of cycle 16. Therefore, the temperature of the adsorbent pellets is high and becomes constant at $t = 3606$ min in cycle 16. The temperature decreases at $t = 3900$ min in the adsorption process of cycle 17 and at $t = 4192$ min in the adsorption process of cycle 18 because the engine is stopped for approximately half a day. Consequently, the appropriate temperatures are achieved for both the adsorption and desorption processes.

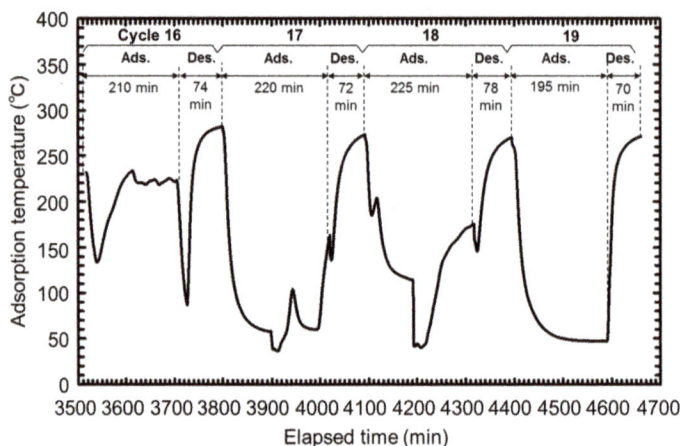

Figure 8. Time-dependent temperature in the adsorption chamber packed with 80 kg of adsorbent pellets in cycles 16–19.

Table 4 shows the resulting adsorbed, desorbed, reduced, and treated amounts of NO_x in cycles 16–19, as well as the gaseous flow rates and energy efficiencies in aftertreatment. The adsorbed mass of NO_x, W_a, ranges from 855 to 1651 g(NO_2). The desorbed mass of NO_x, W_d, ranges from 41.4 to 160 g(NO_2). The mass of NO_x reduced by the application of NTP, W_{NTP}, is in the range of 17.3–114 g(NO_2). The total amount of NO_x removed by the system is calculated as

$$W_{system} = W_a - W_d + W_{NTP} \tag{2}$$

The energy efficiency of the NTP treatment, which shows how to efficiently treat a mass of NO_x per unit of energy, is calculated as follows.

$$\eta_{NTP} = \frac{W_{NTP}}{E_{NTP}} \tag{3}$$

where, E_{NTP} represents the applied NTP energy. η_{NTP} is determined to be 1.1–8.1 g(NO_2)/kWh. The NO_x removal energy efficiency of the system is calculated as

$$\eta_{system} = \frac{W_{system}}{E_{NTP}} \tag{4}$$

The present technology exhibits the highest system energy efficiency, i.e., $\eta_{system} = 115$ g(NO_2)/kWh, in cycle 19. The low NTP power of 12.0 kW contributes to this high efficiency. In the adsorption process of cycle 19, the typical concentrations of gaseous NO_2, NO, CO, and O_2 downstream of the adsorption chamber are 100 ppm, 430 ppm, 69 ppm, and 13.9%, respectively. In the desorption process of cycle 19, the NO_x concentrations upstream and downstream of the NTP generator are 5610 and 1620 ppm, respectively.

Figure 9 shows the relationship between the mass of desorbed NO_x and the reduction energy efficiency in the NTP treatment in the desorption processes of cycles 16–19. The data for the desorption processes of cycles 5–12 of the previous experiments [2] are also shown. Cycles 13–15 are not shown, because NO_x reduction via NTP is not performed. The data plots are presented with the time period of the desorption process. The relationship between the reduction and the mass of desorbed NO_x is approximately given by the line of

$$\eta_{NTP} = 0.0442 W_d. \tag{5}$$

The coefficient, 0.0442, is improved compared with that reported in the previous study [2], because it is determined using a larger amount of data in repeated experiments. Furthermore, a high reduction efficiency of 71% is achieved in cycle 19 for the discharge power of 12.0 kW. The efficiency of reduction via NTP, η_{re}, is defined as the ratio of the amount of reduced NO_x to the amount of desorbed NO_x:

$$\eta_{re} = W_{NTP}/W_d \times 100. \tag{6}$$

Table 4. Treated NO_x amount and removal energy efficiency $(g(NO_2)/kWh)$ in cycles 16–19.

Cycle	16	17	18	19
Averaged flow rate of exhaust gas, Nm^3/h	944	999	948	946
(1) Adsorbed, W_a, $g(NO_2)$	1248	855	937	1651
(2) Desorbed, W_d, $g(NO_2)$	49.9	113	41.4	160
(3) Reduced by NTP, W_{NTP}, $g(NO_2)$	26.0	35.4	17.3	114
(4) Removed in system, W_{system}, $g(NO_2)$ $W_{system} = W_a - W_d + W_{NTP}$	1224	777	913	1605
NTP power, kW	12.0	12.0	12.0	12.0
NTP energy, E_{NTP}, kWh	14.8	14.4	15.6	14.0
$\eta_{NTP} = W_{NTP}/E_{NTP}$, $g(NO_2)/kWh$	1.8	2.5	1.1	8.1
$\eta_{system} = W_{system}/E_{NTP}$, $g(NO_2)/kWh$	82.7	54.0	58.5	115

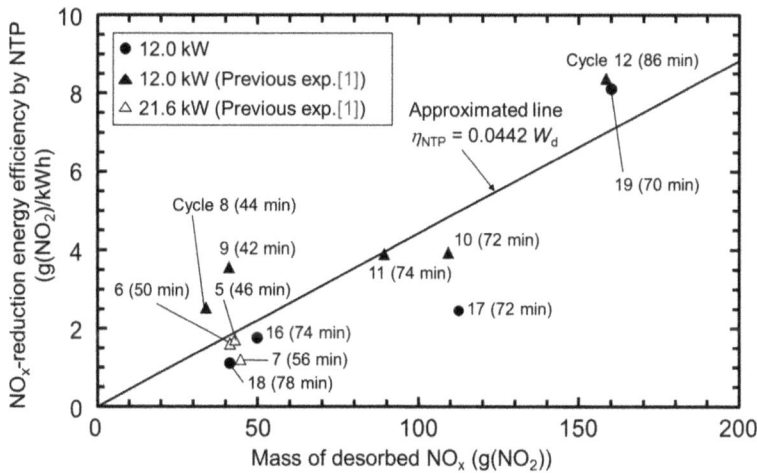

Figure 9. Relationship between the energy efficiency of NO_x reduction via NTP and the mass of desorbed NO_x from the adsorbent in the adsorption process with 80 kg of adsorbent pellets.

The system energy efficiency of $\eta_{system} = 115\ g(NO_2)/kWh$ is lower than $\eta_{system} = 161\ g(NO_2)/kWh$ observed in the previous study [2]. This is because the previous investigation is performed by exploiting the high-adsorption performance of relatively new adsorbents. However, the present study is conducted in the steady state of the adsorption and desorption of NO_x, in which the adsorption performance decreases. However, the desorption performance and efficiency of NO_x reduction via NTP are higher those in the previous study. For a marine diesel engine with a rotation speed of 900 rpm, NO_x emissions should be reduced by 6.89 g/kWh to satisfy the IMO emission standards from Tier II to

III. The recorded energy efficiency of the system (η_{system} = 115 g(NO$_2$)/kWh) corresponds to only 6.0% (6.89/115 × 100) of the engine output power satisfying the requirement. Thus, the high-performance aftertreatment using the present technology satisfies the most recent IMO emission standards.

5. Conclusions

A pilot-scale aftertreatment technology for NO$_x$ reduction in marine diesel exhaust gas was developed. An experiment using a marine diesel engine (output power of 1 MW) was conducted using an NTP generator with a power of 12.0 kW for a larger number of cycles compared with our previous study. The amount of adsorbents was 80 kg. The characteristics of NO$_x$ adsorption/desorption and the NO$_x$ reduction efficiencies were analyzed according to experimental data. The experiments were repeated for up to 19 cycles (longer period than the previous study). Significantly more data were obtained to increase the accuracy for estimating the efficiency of NO$_x$ reduction via NTP. A high reduction efficiency of 71% was achieved using NTP. Additionally, the technology exhibited a high system energy efficiency of 115 g(NO$_2$)/kWh for NO$_x$ removal. Given that a high-concentration NO$_x$ was treated by NTP after NO$_x$ adsorption and desorption from adsorbents, the present aftertreatment can simultaneously achieve high reduction and energy efficiencies. This high efficiency satisfies the most recent requirement of NO$_x$ reduction of 6.89 g/kWh based on the IMO emission standards from Tier II to III for a marine diesel engine with a rotation speed of 900 rpm [2]. Thus, the efficient aftertreatment technology requires only 6.0% of the engine output power. The present aftertreatment technology can satisfy the same-level requirement in the future. Because this aftertreatment technology does not use any rare or precious-metal catalysts, harmful ammonia, or a urea-solution storage tank inside the ship, it has significant advantages compared with conventional exhaust-gas treatments, such as the marine SCR method.

Author Contributions: Conceptualization, K.S. and M.O.; Methodology, T.K. (Takuya Kuwahara), K.Y., T.K. (Tomoyuki Kuroki), K.H., and M.O.; investigation, T.K. (Takuya Kuwahara), K.Y., T.K. (Tomoyuki Kuroki), K.H., and M.O.; Resources, K.S. and M.O.; Writing—original draft preparation, T.K. (Takuya Kuwahara) and M.O.; Writing—review and editing, T.K. (Takuya Kuwahara), K.Y., T.K. (Tomoyuki Kuroki), K.H., K.S. and M.O.; Funding acquisition, M.O.

Funding: This work was supported by JSPS KAKENHI grant numbers 24246145 and 249848.

Acknowledgments: The authors are grateful to M. Nishimoto, M. Kawai, T. Shinohara (former students at Osaka Prefecture University), and S. Tagawa (Nara Prefectural Institute of Industrial Development) for their contributions to the study.

Conflicts of Interest: The authors declare no conflicts of interest.

References

1. Okubo, M.; Kuwahara, T. *New Technologies for Emission Control in Marine Diesel Engines*; Elsevier: Oxford, UK, 2019; ISBN 978-0-12-812307-2.
2. Kuwahara, T.; Yoshida, K.; Kuroki, T.; Hanamoto, K.; Sato, K.; Okubo, M. Pilot-Scale Aftertreatment Using Nonthermal Plasma Reduction of Adsorbed NO$_x$ in Marine Diesel-Engine Exhaust Gas. *Plasma Chem. Plasma Process.* **2014**, *34*, 65–81. [CrossRef]
3. Kuwahara, T.; Yoshida, K.; Hanamoto, K.; Sato, K.; Kuroki, T.; Okubo, M. A Pilot-Scale Experiment for Total Marine Diesel Emission Control Using Ozone Injection and Nonthermal Plasma Reduction. *IEEE Trans. Ind. Appl.* **2015**, *51*, 1168–1178. [CrossRef]
4. Grados, C.V.D.; Uriondo, Z.; Clemente, M.; Espadafor, F.J.J.; Gutiérrez, J.M. Correcting Injection Pressure Maladjustments to Reduce NO$_x$ Emissions by Marine Diesel Engines. *Transp. Res. Part. D Transp. Environ.* **2009**, *14*, 61–66. [CrossRef]
5. Forzatti, P. Present Status and Perspectives in De-NO$_x$ SCR Catalysis. *Appl. Catal. A Gen.* **2001**, *222*, 221–236. [CrossRef]
6. Kang, M.; Park, E.D.; Kim, J.M.; Yie, J.E. Cu–Mn Mixed Oxides for Low Temperature NO Reduction with NH3. *Catal. Today* **2006**, *111*, 236–241. [CrossRef]

7. Gómez-García, M.A.; Zimmermann, Y.; Pitchon, V.; Kiennemann, A. Multifunctional Catalyst for De-NO$_x$ Processes: The Selective Reduction of NO$_x$ by Methane. *Catal. Commun.* **2007**, *8*, 400–404. [CrossRef]
8. Cheng, X.; Bi, X.T. A Review of Recent Advances in Selective Catalytic NO$_x$ Reduction Reactor Technologies. *Particuology* **2014**, *16*, 1–18. [CrossRef]
9. Kwon, D.W.; Nam, K.B.; Hong, S.C. The Role of Ceria on the Activity and SO$_2$ Resistance of Catalysts for the Selective Catalytic Reduction of NO$_x$ by NH$_3$. *Appl. Catal. B Environ.* **2015**, *166*, 37–44. [CrossRef]
10. Boscarato, I.; Hickey, N.; Kašpar, J.; Prati, M.V.; Mariani, A. Green Shipping: Marine Engine Pollution Abatement Using a Combined Catalyst/Seawater Scrubber System. 1. Effect of Catalyst. *J. Catal.* **2015**, *328*, 248–257. [CrossRef]
11. Da Cunha, B.N.; Gonçalves, A.M.; da Silveira, R.G.; Urquieta-González, E.A.; Nunes, L.M. The Influence of a Silica Pillar in Lamellar Tetratitanate for Selective Catalytic Reduction of NO$_x$ Using NH$_3$. *Mater. Res. Bull.* **2015**, *61*, 124–129. [CrossRef]
12. McAdams, R.; Beech, P.; Shawcross, J.T. Low Temperature Plasma Assisted Catalytic Reduction of NO$_x$ in Simulated Marine Diesel Exhaust. *Plasma Chem. Plasma Process.* **2008**, *28*, 159–171. [CrossRef]
13. Wang, H.; Li, X.; Chen, P.; Chen, M.; Zheng, X. An Enhanced Plasma-Catalytic Method for DeNO$_x$ in Simulated Flue Gas at Room Temperature. *Chem. Commun.* **2013**, *49*, 9353–9355. [CrossRef] [PubMed]
14. Jõgi, I.; Stamate, E.; Irimiea, C.; Schmidt, M.; Brandenburg, R.; Hołub, M.; Bonisławski, M.; Jakubowski, T.; Kääriäinen, M.-L.; Cameron, D.C. Comparison of Direct and Indirect Plasma Oxidation of NO Combined with Oxidation by Catalyst. *Fuel* **2015**, *144*, 137–144. [CrossRef]
15. Pietikäinen, M.; Väliheikki, A.; Oravisjärvi, K.; Kolli, T.; Huuhtanen, M.; Niemi, S.; Virtanen, S.; Karhu, T.; Keiski, R.L. Particle and NO$_x$ Emissions of a Non-Road Diesel Engine with an SCR Unit: The Effect of Fuel. *Renew. Energy* **2015**, *77*, 377–385. [CrossRef]
16. Guo, M.; Fu, Z.; Ma, D.; Ji, N.; Song, C.; Liu, Q. A Short Review of Treatment Methods of Marine Diesel Engine Exhaust Gases. *Procedia Eng.* **2015**, *121*, 938–943. [CrossRef]
17. Manivannan, N.; Agozzino, G.; Balachandran, W.; Abbod, M.F.; Jayamurthy, M.; Natale, F.D.; Brennen, D. NO Abatement Using Microwave Micro Plasma Generated with Granular Activated Carbon. *IEEE Trans. Ind. Appl.* **2017**, *53*, 5845–5851. [CrossRef]
18. Madhukar, A.; Rajanikanth, B.S. Augmenting NO$_x$ Reduction in Diesel Exhaust by Combined Plasma/Ozone Injection Technique: A Laboratory Investigation. *High. Volt.* **2018**, *3*, 60–66. [CrossRef]
19. Hołub, M.; Kalisiak, S.; Borkowski, T.; Myśków, J.; Brandenburg, R. The Influence of Direct Non-Thermal Plasma Treatment on Particulate Matter (PM) and NO$_x$ in the Exhaust of Marine Diesel Engines. *Pol. J. Environ. Stud.* **2010**, *19*, 1199–1211.
20. Hołub, M.; Borkowski, T.; Jakubowski, T.; Kalisiak, S.; Myśków, J. Experimental Results of a Combined DBD Reactor-Catalyst Assembly for a Direct Marine Diesel-Engine Exhaust Treatment. *IEEE Trans. Plasma Sci.* **2013**, *41*, 1562–1569. [CrossRef]
21. Schmidt, M.; Basner, R.; Brandenburg, R. Hydrocarbon Assisted NO Oxidation with Non-Thermal Plasma in Simulated Marine Diesel Exhaust Gases. *Plasma Chem. Plasma Process.* **2013**, *33*, 323–335. [CrossRef]
22. Seddiek, I.S.; Elgohary, M.M. Eco-Friendly Selection of Ship Emissions Reduction Strategies with Emphasis on SO$_x$ and NO$_x$ Emissions. *International J. Nav. Archit. Ocean. Eng.* **2014**, *6*, 737–748. [CrossRef]
23. Han, Z.; Liu, B.; Yang, S.; Pan, X.; Yan, Z. NO$_x$ Removal from Simulated Marine Exhaust Gas by Wet Scrubbing Using NaClO Solution. *J. Chem.* **2017**, *2017*, 9340856. [CrossRef]
24. Yoshida, K.; Kuroki, T.; Okubo, M. Diesel Emission Control System Using Combined Process of Nonthermal Plasma and Exhaust Gas Components' Recirculation. *Thin Solid Films* **2009**, *518*, 987–992. [CrossRef]
25. Kuwahara, T.; Yoshida, K.; Kannaka, Y.; Kuroki, T.; Okubo, M. Improvement of NO$_x$ Reduction Efficiency in Diesel Emission Control Using Nonthermal Plasma Combined Exhaust Gas Recirculation Process. *IEEE Trans. Ind. Appl.* **2011**, *47*, 2359–2366. [CrossRef]
26. Okubo, M.; Tanioka, G.; Kuroki, T.; Yamamoto, T. NO$_x$ Concentration Using Adsorption and Nonthermal Plasma Desorption. *IEEE Trans. Ind. Appl.* **2002**, *38*, 1196–1203. [CrossRef]
27. Okubo, M.; Inoue, M.; Kuroki, T.; Yamamoto, T. NO$_x$ Reduction after Treatment System Using Nitrogen Nonthermal Plasma Desorption. *IEEE Trans. Ind. Appl.* **2005**, *41*, 891–899. [CrossRef]
28. Yoshida, K.; Okubo, M.; Yamamoto, T. Distinction between Nonthermal Plasma and Thermal Desorptions for NO$_x$ and CO$_2$. *Appl. Phys. Lett.* **2007**, *90*, 131501. [CrossRef]

29. Yoshida, K.; Okubo, M.; Kuroki, T.; Yamamoto, T. NO$_x$ Aftertreatment Using Thermal Desorption and Nitrogen Nonthermal Plasma Reduction. *IEEE Trans. Ind. Appl.* **2008**, *44*, 1403–1409. [CrossRef]
30. Okubo, M.; Yamamoto, Y.; Kuroki, K. Exhaust Gas Cleaning Method and System. Japan Patent No. 2003-361010, 21 October 2003.
31. Kuwahara, T.; Yoshida, K.; Hanamoto, K.; Sato, K.; Kuroki, T.; Yamamoto, T.; Okubo, M. Pilot-Scale Experiments of Continuous Regeneration of Ceramic Diesel Particulate Filter in Marine Diesel Engine Using Nonthermal Plasma-Induced Radicals. *IEEE Trans. Ind. Appl.* **2012**, *48*, 1649–1656. [CrossRef]

MDPI

St. Alban-Anlage 66

4052 Basel

Switzerland

Tel. +41 61 683 77 34

Fax +41 61 302 89 18

www.mdpi.com

Energies Editorial Office

E-mail: energies@mdpi.com

www.mdpi.com/journal/energies

www.ingramcontent.com/pod-product-compliance
Lightning Source LLC
Chambersburg PA
CBHW051914210326
41597CB00033B/6135